THE
QUANTUM ENIGMA

Finding the Hidden Key

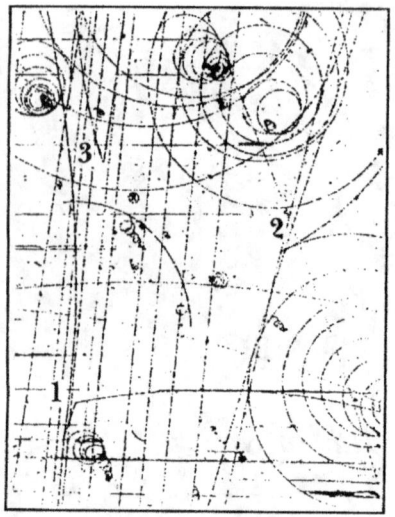 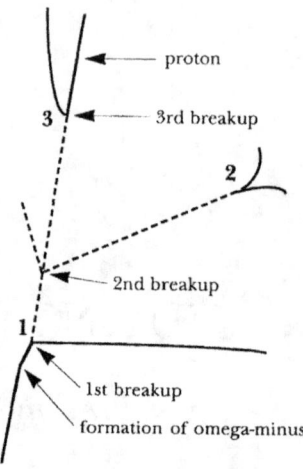

proton

3rd breakup

2nd breakup

1st breakup

formation of omega-minus

The bubble-chamber photograph on the front cover is reproduced with the permission of the Brookhaven National Laboratory. Taken in 1964, it marks the discovery of omega-minus, a heavy particle predicted by Murray Gell-Mann. This photograph provided the first experimental confirmation of Gell-Mann's theory of quarks, for which he received the Nobel Prize in 1969. It depicts the formation and almost immediate disintegration of an omega-minus particle. The disintegration is accomplished in three successive breakups, in each of which the particle sheds a dose of what physicists term 'strangeness'. The end product of this sequence is a proton (a particle having zero strangeness).

The diagram indicates the formation and successive breakup points of the omega-minus particle. Dashed lines represent trajectories of neutral particles (which leave no tracks in a bubble chamber). The remaining trajectories, on the other hand, can be readily identified. Points 1, 2, and 3 have been marked on the accompanying photograph.

THE
QUANTUM ENIGMA

Finding the Hidden Key

FOURTH EDITION

Wolfgang Smith

Philos-Sophia Initiative

Fourth, PSI edition, 2023
Third, revised edition, 2005
First, and Second, revised editions,
Sherwood Sugden & Co., 1995

To request permission, contact the publisher at
info@philossophiainitiative.com

Hardcover ISBN: 979-8-9885769-0-7
Paperback ISBN: 979-8-9885769-1-4
eBook ISBN: 979-8-9885769-2-1

Library of Congress Control Number: 2023911099

Philos-Sophia Initiative
www.philos-sophia.org

CONTENTS

FOREWORD

SINCE THE BEGINNING of the development of quantum mechanics, different interpretations have been given as to its meaning not only by physicists, especially Bohr and Einstein, but also by a number of philosophers. These interpretations have met with little success, however, in providing intelligibility for the consequences of what is observed and measured on the experimental level. The present work is the first by a qualified scientist to bring to bear not a rationalistic or empirical philosophy but traditional metaphysics, ontology, and cosmology upon quantum mechanics in order to provide the key for the understanding of the real significance of this basic physical science. The result is one of the most important books to appear on the explanation of modern physics in the light of the eternal truths of the perennial philosophy and on the categorical refutation of the scientism and reductionism that characterizes so much of the current understanding of modern science.

The author, who is well known to students in the field of the relation between religion and science through his early works *Cosmos and Transcendence* and *Teilhardism and the New Religion*, as well as his recent *The Wisdom of Ancient Cosmology*, is deeply grounded in traditional metaphysics and theology, especially the school of St Thomas Aquinas. At the same time he is a notable scientist well versed in the intricacies of quantum mechanics. He therefore writes with an authority that shines through the pages of his book, providing a treatment of the subject that stands at the antipodes of the genre of shallow syntheses between modern physics and oriental metaphysics so common today, and espoused especially by certain currents of what is now referred to as 'The New Religions'.

In his preface the author points out that there is in fact no consistent quantum mechanical worldview, despite the remarkable accuracy of predictions on the basis of its theories and models. The situation has become so difficult that it has prompted one recent author to speak of a 'reality marketplace'. It is to discover the authentic worldview to which quantum mechanics points, and that can make possible an intelligible understanding of it, that Prof. Smith set out to write this book.

Also provided is a philosophy of quantum mechanics drawn from traditional ontological, cosmological, and metaphysical doctrines, replacing the prevalent Cartesianism still underlying modern science despite the many changes quantum mechanics has brought about. By doing so he has removed the contradictions apparent in quantum mechanics as viewed ordinarily, and has made the subject intelligible from the point of view of the *philosophia perennis*. His clear distinction between the physical and corporeal, which is one of the main contributions of the book, has situated the ontological status of the subject matter of modern physics in the universal hierarchy of being. He has also freed the prevalent understanding of the corporeal world and the qualitative sciences associated with it over the ages from the stranglehold of a purely quantitative science, and has destroyed once and for all scientistic reductionism, which is one of the pillars of the modern and postmodern worldviews.

The Quantum Enigma is of great significance not only for the philosophy of science, but also for the whole domain of human knowledge, and should be disseminated as widely as possible. It marks the first encounter in depth between traditional ontology and quantum mechanics in the mind of a person who is a master in both domains, and is able to provide a metaphysical understanding of modern physics, its achievements and limitations. It is in fact a counterweight to so many works which move in the other direction by interpreting millennial metaphysical teachings of East and West in light of modern physics. The book is clearly written, the technical mathematical treatment of quantum mechanics being confined to an appendix which can be consulted by those with the necessary background. The work itself, however, does not require technical knowledge of mathematical physics but addresses all those who seek an understanding of the world about them and the meaning that modern science has in both explaining a particular dimension of this world and veiling its qualitative aspects from modern man. All those overwhelmed and distraught by reductionism, scientism, and the excessive pretensions of a purely quantitative science, and who are at the same time aware of both the achievements and ambiguities of quantum mechanics, will be grateful to Wolfgang Smith for having produced a work of exceptional significance in destroying the extravagant claims of scientism, and yet unravelling at the same time the enigma of quantum mechanics in light of perennial doctrines which have always provided the means for solving the enigmas and riddles of human existence and thought over the ages.

SEYYED HOSSEIN NASR

GEORGE WASHINGTON UNIVERSITY

PREFACE
TO THIRD EDITION

IN THE DECADE that has elapsed since the First Edition appeared, the basic conceptions introduced in this book have proved to be seminal in regard to a broad range of cosmological issues. One of the most direct applications, pertaining to the domain of astrophysics, has brought to light the categorical limitations of contemporary cosmologies. It appears that the ontological lessons learned in the context of quantum theory prove to be decisive in the astrophysical realm as well. Another quite immediate application of the same principles to what is currently termed anthropic coincidence has radically altered the picture: the phenomena in question turn out not to be 'coincidental' at all, but are in fact logically implied on ontological grounds. In regard to cosmography—to mention a third line of inquiry—one finds that the apparent conflict between modern science and ancient 'world pictures' is by no means as absolute as one tends to imagine: the same 'ontological lessons learned' entail that it is not the so-called hard facts of science that rule out alternative cosmographies, but that the stipulated incompatibility derives in fact from presuppositions of a Cartesian kind. One is amazed to see how different the world looks when it is no longer viewed through Cartesian spectacles. As a rule one discovers that once the ontological muddle underlying contemporary scientific thought has been exposed and eliminated, the way is clear to an integration of actual scientific findings into orders of knowledge pertaining to what has sometimes been termed the perennial wisdom of mankind.[1]

---

1. Ten studies of this kind have been published in my recent book, *Ancient Wisdom and Modern Misconceptions* (Tacoma, WA: Angelico Press/Sophia Perennis, 2013).

Among the ideas introduced in the context of quantum theory which have found application beyond physics, the concept of 'vertical causality', defined in Chapter 6, deserves to be singled out on account of its intimate connection with a new and increasingly influential domain of science known as the theory of intelligent design. The central result of ID theory is a theorem to the effect that a quantity termed complex specified information cannot be increased by any temporal process, be it deterministic, random or stochastic.[2] This means, in light of our analysis, that vertical causation alone can give rise to CSI. Our main result, to the effect that state vector collapse must likewise be attributed to vertical causality, assumes thus an enhanced significance. Vertical causation, so far from constituting a *deus ex machina* for the resolution of quantum paradox, as critics might charge, constitutes indeed a universal principle of causality which modern science is obliged finally to recognize. It turns out that a multitude of natural phenomena, from the collapse of a state vector to the genesis of biological organisms, demands that hitherto unacknowledged kind of causality. Given the fact that contemporary science, by the very nature of its *modus operandi*, is geared to deal exclusively with temporal or 'horizontal' modes of causation, this implies that the phenomena in question cannot, strictly speaking, be explained or understood in scientific terms: like it or not, metaphysical principles have perforce entered the picture, in defiance of the prevailing naturalism.

The present Third Edition offers a revised version of Chapter 6. In the original presentation the subject of vertical causality was broached by way of astrophysical cosmology, which I had as yet inadequately probed, be it from a scientific or from an ontological point of view. From both directions, meanwhile, difficulties regarding that theory have come into view, of which I had been insufficiently aware at the time of writing the original text.[3] In the new

2. A stochastic process is one that entails both chance and necessity, as is the case, for example, in Brownian motion. For a brief account of ID theory and its connection with vertical causality I refer to *Ancient Wisdom and Modern Misconceptions*, chap. 10.

3. It should be noted that from a logical, and indeed from a symbolist point of view, the connection between the stipulated initial singularity and vertical causation

version of Chapter 6 I have dropped all reference to big bang cosmology and have dealt with the etiological issues directly. After introducing the concept of vertical causality in general terms I proceed to explain its relevance not only to the phenomenon of state vector collapse, but indeed to quantum mechanics at large. What appears incongruous and bizarre from the customary Cartesian angle of vision turns out now to be precisely what is called for ontologically: this is what the train of argument, begun in Chapter 1 and consummated in Chapter 6, brings to light.

PREFACE TO FIRST EDITION

THIS BOOK IS ABOUT QUANTUM PHYSICS, or what has been termed the 'quantum reality' problem. It deals with an enigma that has tantalized physicists, philosophers, and an ever-widening public for decades. The pertinent literature is vast, and it would appear that just about every conceivable avenue of approach to the problem—no matter how seemingly farfetched—has been advocated somewhere and explored. Gone are the days when the authority of physics could be invoked in support of a single established world-view! What has happened is that the pre-quantum scientistic world-view (now termed 'classical') has come to be disavowed 'at the top': by physicists capable of grasping the implications of quantum theory. And this in turn has called forth an abundance of conjectured alternatives, competing with one another, as it were, to fill the ontological void—a situation that has prompted one recent author to speak of a 'reality marketplace.' Quantum mechanics, if you will, is a scientific theory in search of a *Weltanschauung*. The search has been on since 1927.[4]

---

as explained in the original version of Chapter 6 remains valid, regardless whether big bang cosmology proves to be factual.

4. The new physics came to birth during the years 1925 and 1926. By the time physicists gathered at Como in 1927 for the International Physics Conference, the foundations of nonrelativistic quantum theory had been laid. It was later during

Meanwhile the spectacle of a dozen top-ranking scientists promoting twelve different world-views is hardly reassuring; and there is the temptation to conclude that truth is unattainable, or, worse still, that it is relative, a matter simply of personal opinion.

What is called for, however, is a closer look at the foundations of scientific thought: at the hidden assumptions that have conditioned our contemporary intellectual perceptions. A modest probe into matters generally ignored suffices to reveal a startling fact: it happens that every quantum-reality position thus far enunciated hinges upon one and the same ontological presupposition, a tenet which moreover derives from the philosophical speculations of Galileo and Descartes, and which, surprisingly enough, has been sharply and cogently attacked by some of the most eminent philosophers of the twentieth century. It may indeed seem strange that an ontological assumption that has thus become suspect, to say the least, should have remained unchallenged throughout the length and breadth of the quantum reality debate;[5] but one must remember that the notion of which we speak has become ingrained in the scientific mentality to the point where it can hardly be recognized as a presupposition, let alone as a spurious premise that must go.

My fundamental claim can now be stated quite simply: Remove this error, expose this virtually ubiquitous assumption as the fallacy it is, and the pieces of the quantum puzzle begin to fall into place. The very features of quantum theory, in fact, which, prior to this ontological rectification had seemed the most incomprehensible, prove now to be the most enlightening. As might be surmised, these features bear witness, on a technical level, to an ontological fact, a truth which had hitherto been obscured.

---

the same year, when physicists met again in Brussels for the Fifth Solvay Conference, that the quantum debate erupted in full force, so to speak, in the form of the celebrated Bohr-Einstein exchange.

5. The lone exception appears to be the case of Werner Heisenberg. But whereas Heisenberg has occasionally questioned the offending premise, and has gone so far as to suggest that it may be the main cause of incomprehension among physicists, his own interpretation of quantum theory, as we shall see, presupposes this tenet nonetheless.

My first major objective will be to identify this elusive and fallacious premise, and refute it with optimum cogency. Following this, I shall need to give a revised account of the *modus operandi* by which physics is defined, an account which no longer hinges upon the now disqualified axiom. This done, we shall be in a position to reflect anew upon the salient findings of quantum theory, to see whether these strange and puzzling facts can at last be understood. And this is the task which will occupy the remainder of the book.

At the top of the list of 'strange facts' that demand an explanation stands the phenomenon of state vector collapse, which could well be termed the central enigma of quantum physics. It poses a fundamental problem that cannot be ignored or by-passed if one would understand the nature of the physical universe, and its relation to whatever other ontological planes there be.

Considerations of this kind, meanwhile, need not detain the working physicist, nor do they alter the fact that quantum mechanics is beyond doubt the most accurate, the most universal, as well as the most sophisticated scientific theory ever advanced by man. In a thousand hair-splitting experiments it has never yet been proved wrong. But quantum theory does more than answer a multitude of questions: it also raises a few of its own. And whereas classical physics, which by comparison is both crude and inaccurate, generally inspires dreams of omniscience, the new physics counsels caution and a becoming sobriety.

Following upon these cursory observations, let it be said emphatically that the present book is written as much for the general or 'non-mathematical' reader as for the interested physicist. I have taken pains not to presuppose any technical knowledge of physics, or any previous acquaintance with the quantum-reality literature. The requisite technical concepts of quantum theory will be briefly explained in suitably simplified terms. Such unexplained technical notions or references as remain are invariably extraneous to the main argument and should cause no concern to the general reader. For readers with an interest in mathematics I have appended a brief introduction to quantum theory, which provides a glimpse of its mathematical structure.

I have occasionally used philosophical terms which may not be familiar, and I have been forced, here and there, to coin a few technical expressions of my own. In each case I have done my best to explain the meaning of these special words at the place where they are first introduced. Brief definitions have also been provided in a Glossary.

It needs, finally, to be emphasized that despite its seemingly 'specialized' nature, the quantum-reality problem is beyond doubt the most universally significant question hard science has ever posed. What it demands, clearly, is an integral world-view that breaks radically with the accustomed, the 'classical'; and that is what I propose to supply in the sequel. I will not, however, attempt to preview the conclusions of the inquiry in these prefatory remarks. As concerns the requisite ontological conceptions, these will be unfolded within the context of the quantum reality problematic, each in its place.

I

REDISCOVERING
THE CORPOREAL WORLD

THE DIFFICULTIES AND INDEED PERPLEXITIES which beset us the moment we try to make philosophic sense out of the findings of quantum theory are caused, not just by the complexity and subtlety of the microworld, but first and foremost by an adhesion to certain false metaphysical premises, which have occupied a position of intellectual dominance since the time of René Descartes.

What are these premises? To begin with, there is the Cartesian conception of an external world made up exclusively of so-called *res extensae* or 'extended entities,' concerning which one assumes that they are bereft of all qualitative or 'secondary' attributes, such as color, for instance. All else is relegated, according to this philosophy, to the so-called *res cogitantes* or 'thinking entities,' whose constitutive act, so to speak, is not extension, but thought. Thus, according to Descartes, whatever in the universe is not a *res extensa* is therefore 'an object of thought,' as we would say, or in other words, a thing that has no existence outside of a particular *res cogitans* or mind.

Admittedly the dichotomy has its use; for indeed, by relegating the so-called secondary attributes to the second of the Cartesian compartments, one has at one stroke achieved an incalculable simplification of the first. What remains, in fact, is precisely the kind of 'external world' that mathematical physics could in principle comprehend 'without residue.' There is however a price to be paid: for once the real has been split in two, no one apparently knows how to put the pieces back together again. How, in particular, does *res cogitans* gain knowledge of *res extensa*? By perception, to be sure; but then, what is it that we perceive? Now, in pre-Cartesian days it was generally thought—by philosopher and non-philosopher alike—that in the act of visual perception, for example, we do indeed 'look out upon the external world.' Not so, declares René Descartes; and with good reason, given that one has accepted the Cartesian dichotomy. For if, what I actually perceive, is a red object, let us say, then it must *ipso facto* belong to *res cogitans*, for the simple reason that *res extensa* has no color at all. Thus, following upon his initial assumptions, it was not by choice, but by force of logical necessity that Descartes was led to postulate what has since become known as 'bifurcation': the thesis, namely, that the perceptual object belongs exclusively to *res cogitans*, or that what we actually perceive, in other words, is private and subjective. In crass opposition to common belief, Cartesianism insists that we do *not* 'look out upon the external world'; according to this philosophy we are in reality cooped up, each in his own private world, and what we normally take to be a part of the external universe is in truth but a phantasm, a mental object—like a dream—whose existence does not extend beyond the perceptual act.

But this position is precarious to say the least; for if the act of perception does not in fact span the gap between the inner and the outer worlds—between *res cogitans* and *res extensa*—how then is the gap bridged? How, in other words, is it possible to know external things, or even to know that there exists an external world in the first place? Descartes himself, as one will recall, experienced great difficulty in overcoming his celebrated doubts, and was able to do so only by way of a tortuous argument which few today would find convincing. Is it not strange that tough-minded scientists should have so readily, and for so long, espoused a rationalist doctrine which calls in question the very possibility of empirical knowledge?

But then, if one ignores this epistemological impasse—or pretends that it has been resolved—one is able to derive satisfaction from the apparent benefit which Cartesianism does confer: for as I have already pointed out, the simplification of the external world resulting from bifurcation renders thinkable a mathematical physics of unlimited scope. But the question, in any case, is not whether bifurcation is in some sense advantageous, but simply whether it is true and indeed tenable. And that is the issue that needs in first place to be resolved; all other matters pertaining to the interpretation of physics are obviously consequent upon this, and must therefore await their turn.

❖   ❖   ❖

Prior to science, prior to philosophy, prior to every ratiocinative inquiry, the world exists and is known in part. It exists not necessarily in the specific sense in which certain scientists or philosophers may have imagined that it does or does not, but precisely as something that can and must on occasion present itself to our inspection. It must so present itself, moreover, by a kind of logical necessity, for it belongs to the very conception of a world to be partially known—even as it belongs to the nature of a circle to enclose some region of the plane. Or to put it another way: If the world were *not* known in part, it would *ipso facto* cease to be the world—'our' world, in any case. Thus in a sense—which can however be easily misconstrued!—the world exists 'for us'; it is there 'for our inspection,' as I have said.

Now that inspection, to be sure, is accomplished by way of our senses, by way of perception; only it is to be understood from the start that perception is not sensation, pure and simple, which is to say that it is not just a passive reception of images, or an act bereft of human intelligence. But regardless of how the act is consummated, the fact remains that we do perceive the things that surround us; circumstances permitting, we can see, touch, hear, taste and smell them, as everyone knows full well.

It is thus futile and perfectly vacuous to speak of the world as something that is in principle unperceived and unperceivable; and besides, it is an offense against language—much like saying that the ocean is dry, or a forest void. For the world is manifestly conceived as the locus of things perceptible; it consists of things, which though they may not now be actually perceived, *could* nonetheless be perceived under suitable conditions: that is the crux of the matter. For example, I now perceive my desk (through the senses of sight and touch); and when I leave my study, I will no longer perceive it; but the point is of course that upon my return I can perceive it again. As

Bishop Berkeley had correctly observed, to say that a corporeal object exists is to say, not that it *is* perceived, but that it can and will be perceived under appropriate circumstances.

It is this vital and oft-forgotten truth that underlies his justly celebrated dictum '*Esse est percipi*' ('To be is to be perceived'), notwithstanding the fact that this highly elliptical statement can indeed be interpreted in the sense of a spurious idealism. This danger—to which the Irish bishop himself fell prey[1]—arises, moreover, mainly from the circumstance that the *percipi* in Berkeley's formula can readily be misconceived. As I have already pointed out, perception can be misconstrued as a mere sensation; and that is essentially what most philosophers took it to be, from the time of John Locke right up to the twentieth century, when it came to pass that this crude and insufficient view was subjected to scrutiny and discarded by the leading schools.

❖   ❖   ❖

Granted that we do perceive the external object, it is of course to be admitted that we are able to perceive it only in part, and that the bulk of the entity, so to speak, remains perforce hidden from view. Thus, in the paramount case, which is evidently that of visual perception, it is normally the outer surface that is visible, while the interior remains unperceived. Now it may seem to some that in order to perceive an object one would have to perceive it in full—a fact which would obviously imply that we can never perceive anything at all. But then, does not the circumstance that we do perceive only in part militate in reality, not against the supposition that we perceive external objects, but against the 'all or nothing' view of perception, precisely?

The fact is that it belongs to the very nature of the object to manifest itself only in part, even as it belongs to a circle, let us say, to exclude an indefinite portion of the plane. There is a simple and

---

1. I have discussed the philosophies of Descartes, Berkeley and Kant apropos of bifurcation in *Cosmos and Transcendence* (Philos-Sophia Initiative, 2021), chap. 2.

obvious 'principle of indetermination,' operative within the familiar corporeal domain, which affirms that neither the external world at large, nor the least object therein, can be known or perceived 'without residue.' It cannot be thus known, moreover, not simply or unilaterally on account of a certain incapacity on the part of the human observer, but also by the very nature of the corporeal entity itself. It is of course always possible to perceive more, and thus to extend our perceptual knowledge, even as it is possible to enlarge a circle; what is not possible, on the other hand, is to 'exhaust' the object by way of perception—to enlarge the circle to the point where it ceases to exclude some 'infinite remnant' of the plane. For it is to be noted that a corporeal object 'fully perceived' would cease to be a corporeal object, even as a circle 'without exterior' would cease to be a circle.

To put it simply: If we could 'look out upon the world' with the Eye of God, the world as such would forthwith cease to exist—even as the pictures on a cinema screen would disappear the moment a sufficiently bright light is switched on.

The cinematic metaphor, of course, must not be pressed too far; for if God does 'see' the corporeal world, this 'perception' obviously does not obliterate the contents of that world. But even though corporeal entities remain, they are not what an omniscient observer would himself behold, the point being, once again, that a corporeal object 'fully known' would *ipso facto* cease to be a corporeal object. We must bear in mind that these entities—by definition, if you will—exist 'for us' as things to be explored by way of perception.

The fact is that 'we' are somehow present upon the scene—not, in this instance, as an object, but as a subject, precisely. And though this subjective presence can indeed be forgotten or ignored, it cannot be exorcised—which is to say that on closer examination it is bound to show up in the very nature of the object itself. In various ways the object displays of necessity the marks of relativity, of being orientated, so to speak, towards the human observer.

One such 'mark' we have just considered: that it pertains to the object to be known or perceived only in part. However, besides the fact that we perceive only in part, it is likewise plain that what we do perceive is incurably 'contextual.' And this too constitutes an inalienable feature of the object itself. In other words, the attributes of a corporeal object are without exception contextual in a certain sense.

Let us examine the matter. The perceived shape of a body, for example, depends upon our position relative to the object, even as the perceived color depends upon the light in which it is viewed. But whereas the contextuality of shape is generally accepted without a qualm, one is prone to argue that inasmuch as color is a 'contextual attribute,' it must therefore be also a 'secondary attribute' in the Cartesian sense. But why? What actually hinders a contextual attribute from being objectively real? The answer is that nothing so hinders—so long as we entertain a realistic notion of objectivity.

So far as the contextuality of shape is concerned, it is evident that the perceived two-dimensional shapes can be understood as plane projections of an invariant three-dimensional 'shape'; and yet that three-dimensional shape, and indeed all the so-called primary attributes, 'invariant' though they be, are also perforce contextual in a more fundamental sense. An attribute, after all, is nothing more nor less than an observable characteristic of interaction. Mass, for instance, is an observable characteristic of gravitational and inertial interactions; thus we say that a body has so many grams of mass if, when placed on a scale, we observe a corresponding deflection or pointer reading.

In the case of qualitative attributes the principle is the same; color, for instance, is also 'an observable characteristic of interaction'—for as we know, the color of an object is perceived when the latter interacts with a beam of light by reflecting it. There is of course an enormous difference between qualitative and quantitative attributes—a 'categorical' difference, in fact;[2] redness, for example, unlike mass, is not something to be deduced from pointer readings, but something, rather, that is directly perceived. It cannot be quantified, therefore, or entered into a mathematical formula, and consequently cannot be conceived as a mathematical invariant. And yet redness, too, is a kind of invariant; for indeed, if a red object be viewed in white light by an able observer, it will show red—every time!

---

2. Aristotle was wise, after all, when he postulated 'quantity' and 'quality' as separate and irreducible categories.

But not only are both types of attributes incurably contextual, but both alike are objective: color no less than mass. To be objective, after all, is to belong to the object; but what *is* a corporeal object, if not a thing that manifests attributes—both quantitative and qualitative, to be sure—depending upon the conditions in which it is placed. The object, thus, so far from being a Cartesian *res extensa* or a Kantian *Ding an sich*, is in fact conceived or defined in terms of its attributes. To be precise, the concrete object is ideally specified in terms of the full plethora of its attributes; and whereas each of these attributes is in principle observable, it is in the nature of things that most will remain forever unobserved.

What we need above all to understand is that nothing in the world 'simply exists,' but that to exist is precisely to interact with other things—including ultimately observers. The world, therefore, is not to be conceived as a mere juxtaposition of so many individual or self-existent entities—be they *res extensae*, or 'atoms,' or what you will—but must be viewed rather as an organic unity, in which each element exists in relation to every other, and thus in relation to the totality, which includes also, and by force of necessity, a conscious or subjective pole. This fundamental discovery, moreover, which many nowadays associate with recent findings in the domain of quantum physics—or with Eastern mysticism, for that matter—can easily be made 'with the naked eye,' so to speak, for it pertains just as much to the sense-perceived corporeal world as to the newly-discovered quantum domain; it is only that for several centuries we have been prevented from viewing the former without prejudice and distortion caused by erroneous preconceptions of a Cartesian kind.

❖   ❖   ❖

The objection might be raised that quantitative attributes, such as mass, though they be contextual, can nonetheless be conceived as existing in the external world, whereas this is not the case, supposedly, when it comes to a 'perceptual quality' such as redness. It would appear, therefore, that a 'purely objective universe'—a universe, let us say, in which there are no observers at all—can indeed be conceived, but only on condition that it contain no 'secondary attributes' (such as redness).

Let us examine this line of thought. To begin with, one cannot but agree that the idea of a quality, such as redness, bears reference to perception, which is to say that redness is ineluctably something that one perceives. But this does not by any means imply that a thing cannot be red unless it is actually perceived; for we obviously do speak of things unperceived as red, meaning thereby that they *would* show red *if* they were perceived (always, of course, under the stipulation that they are viewed in appropriate light and by a normal or able-bodied observer). The statement that a given object is red is thus conditional, and it is by virtue of this conditionality, precisely, that its truth is independent of whether or not the object is in fact perceived. One may consequently rest assured that a ripe Jonathan, for example, is red even if there be no one in the orchard to perceive it; and if intelligent life on Earth were suddenly to disappear, there is no reason to doubt that the Jonathan would still be red.

There is a sense, then, in which a universe replete with qualitative attributes can be said to exist 'in the absence of human observers'; the real question, therefore, is whether *more* than this could be affirmed with reference to an imagined universe from which all qualities have been deleted. Now, it is of course to be conceded that quantitative attributes, such as mass, for instance, refer less directly to perception—be it visual, tactile, or any other—than color; and this is the reason, presumably, why it may be easier to think of the former as 'primary attributes' in the classic Cartesian sense. But we must not forget that the quantitative attributes with which physics is concerned are after all empirically defined, which is to say that their definition does entail a necessary reference to sense perception, however indirect or remote that reference might be. It is true that the mass of a body is not directly perceived (although the kinesthetic sense may in some instances give us a rough estimate), and that in this regard mass differs from color; but it is also to be noted that the measurement or 'observation' of mass is consummated perforce by a perceptual act. Thus, to say that a body has such and such a mass is to say that a measurement of its mass will give the value in question, which means, once again, that *if* we carry out a certain operation, *then* a corresponding sense perception will ensue (for example, we will perceive this or that number on a scale). The case of mass, therefore, and of the other so-called primary attributes, is not as different from that of color as the Cartesians may think; for in both instances the predication of the attribute (so much mass, or such and such a color) constitutes a conditional statement of exactly the same logical form. A mass no less than a color, therefore, is in a sense a potency to be actualized through an intelligent act involving sense perception. But as a potency each exists in the external world—which is to say that each *exists*, seeing that each *is* a potency. It is all that we can logically ask or reasonably expect of an attribute: to require more would be tantamount to asking that it is and is not actualized at the same time.

So far as objectivity and observer-independence are concerned, therefore, the case for mass and for color stand equally well; both attributes are in fact objective and observer-independent in the strongest conceivable sense. It is only that in the case of mass and other 'scientific' attributes the complexity of the definition makes it easier—psychologically, one might say—to expect the impossible: to forget, in other words, that the world is there 'for us'—as a field to be explored through the exercise of our senses.

❖   ❖   ❖

It may be instructive to reflect upon the fact that there exist 'illusory' perceptions: For example, when we watch a film or a television program, we perceive—or seem to perceive—objects which are not actually present; there are no mountains or rivers within the confines of the theater, nor any men shooting at each other in our living room; and yet we perceive these things as if they were real. Does not this in itself lend support to the bifurcationist contention? Does it not show that what we perceive is in fact subjective—a mere phantasm situated somehow in the brain or mind of the percipient?

Now certainly it proves that what we perceive *may* be subjective, which is to say that there is such a thing as an 'optical illusion' or a false perception. But does it prove that *every* perception is illusory or false? Obviously not. For indeed, the very fact that we speak of an optical illusion, or of a false perception, indicates that there must also exist perceptions which are *not* illusory, not false.

What, then, is the difference between the two cases? The difference, clearly, is that a true or authentic perception satisfies appropriate 'criteria of reality.' If I perceive a river, the question is, can I jump into it? And if I perceive a horse, the question is, can I climb on its back? Thus, with every purported perception of a corporeal entity there is associated a syndrome of 'operational expectations' which can in principle be put to the test; and if (in cases of doubt) some reasonable subset of these have been checked and verified, we then take it that the thing in question is indeed what we have perceived it to be: if I can ride it, hitch it to a wagon, and feed it hay, then it is a horse. And then, of course, my initial perception of the horse was not illusory, but true. Such are the criteria of reality in terms of which one distinguishes between true and false perceptions—and let us not fail to note that the validation of a given perception is accomplished perforce by means of other perceptions, circular as this procedure may seem to the theoretician.

On the other hand, when the bifurcationist informs us that perceptions are 'illusory' (or 'subjective'), he does not mean that they are illusory or false in the normal sense. To the Cartesian philosopher my perceptions of the desk I am writing on are every bit as 'illusory' as is the perception of mountains and rivers in a theater, for both alike are supposedly private phantasms. To be sure, the Cartesian does also distinguish between perceptions that are true or false in the usual sense; he does so by supposing that in the case of a true perception there exists an external object which corresponds to the perceptual in certain specific respects. According to this philosophy, there are in effect two desks: the 'mental' desk which I perceive, and the external desk which I do not perceive. And the two are quite different: the first, for instance, is brown and lacks extension in space, while the second is extended but not brown. But despite these differences the two are supposedly similar in certain respects: If the desk I perceive appears to have a rectangular top, the external desk has a rectangular top as well, and so forth. But all these Cartesian claims are of course purely conjectural, which is to say that it is in principle impossible to ascertain whether any of them are true. More precisely, if the dogma of bifurcation were true, then the corresponding 'two object' theory of perception would *ipso facto* be unverifiable, for the obvious reason that there would be no way of ever finding out whether the external object exists, let alone whether it is geometrically similar to the perceptual. One object is all we ever get to observe, and the stipulation that there are two is perfectly gratuitous. The 'two object' theory of perception, no less than the bifurcationist tenet on which it rests, constitutes thus a metaphysical premise which can be neither verified nor falsified by any empirical or scientific means.

Our question was whether the fact that there are 'illusory' perceptions in the ordinary sense lends support to the bifurcationist contention; and it has now become clear that indeed it does not. The fact that there are optical illusions or hallucinatory perceptions does nothing to show that in the case of an ordinary perception there exist actually two objects as envisioned by the Cartesian philosophy. Indeed, the contrary would seem to be the case: for if an optical illusion or a hallucination is characterized by the fact that the perceptual act miscarries, then this implies that in the case of normal perceptions it does not miscarry; which presumably means that what we then perceive is the external object, precisely.

❖   ❖   ❖

The question arises why Western thought should have been dominated for so long by the Cartesian philosophy, a speculative doctrine which contradicts our most basic intuitions and for which there can in principle be no corroborating evidence. And why should the scientist, of all people, espouse this chimerical teaching, which in effect renders the external world unknowable by empirical means? One might think that he would despise the Cartesian speculation as the idlest of dreams, and of all metaphysical fantasies the most inimical to his purpose. And yet, from the seventeenth century onwards, as we know, Cartesianism and physics have been closely joined, so much so that to the superficial observer it might seem that the dogma of bifurcation constitutes indeed a scientific tenet, supported by all the enormous weight of physical discovery. It was, after all, the great Newton himself who tied the knot of this curious match, and so well, that to the present day the union has proved to be virtually indissoluble.[3]

Yet neither the Cartesian premise nor its association with physics was in fact something altogether new under the sun, for it appears that the first declared bifurcationist in the history of human thought was none other than Democritus of Abdera, the acknowledged father of atomism. 'According to vulgar belief,' declares Democritus, 'there exist color, the sweet and the bitter; but in reality only atoms and the void.'[4] There is a necessary connection, moreover, between the two halves of the doctrine, the point being that he who would explain the universe in terms of 'atoms and the void' must first of all negate the objective reality of the sense-perceived qualities. For as Descartes observed with admirable clarity:

> We can easily conceive how the motion of one body can be caused by that of another, and diversified by the size, figure and situation of its parts, but we are wholly unable to conceive how these same things [size, figure and motion] can produce something else of a nature entirely different from themselves, as for example, those substantial forms and real qualities which many philosophers suppose to be in bodies.[5]

And let us add that even though Descartes does not assume an atomist model of external reality, the difference is quite immaterial as regards the point at issue; for whether one thinks in terms of continuous *res extensae* or in terms of Democritean atoms, the quoted passage suffices in any case to explain why a totalist physics—a physics that would understand the universe 'without residue'—is obliged to accept bifurcation, almost as a 'necessary evil' one might say.

3. See especially E.A. Burtt, *The Metaphysical Foundations of Modern Physical Science* (New York: Humanities Press, 1951).

4. Hermann Diels, *Die Fragmente der Vorsokratiker* (Dublin: Weidmann, 1969), vol. II, p 168.

5. *Principia Philosophiae*, in *Oeuvres* (Paris, 1824,) IV, 198; cited in E.A. Burtt, op. cit., p 112.

It should however be noted that the benefits of bifurcation are more apparent than real; for indeed, the Cartesian is ultimately forced to admit the very thing which 'we are wholly unable to conceive.' He is forced to admit it, namely, when it comes to the process of perception, in which sense-perceived qualities—be they ever so private or 'illusory'—are apparently caused (on the strength of his own assumptions) by 'moving particles.' Like it or not, he is obliged to explain how 'these same things can produce something else of a nature entirely different from themselves,' and must of necessity concede in the end that 'we are wholly unable to conceive' how such a thing is possible. No real philosophic advantage, therefore, results from the postulate of bifurcation, which is to say that the totalist claims of physics need in any case to be relinquished: in a word, not everything without exception can be understood or explained in exclusively quantitative terms.

Getting back to Democritus, it is to be noted that his position was vigorously opposed by Plato and subsequently rejected by the major philosophic schools right up to the advent of modern times; which means that the twin tenets of atomism and bifurcation can indeed be classified as 'heterodox.' But as one also knows, old heresies do not die—they only bide their time, and with the return of conditions favorable to their acceptance they are invariably rediscovered and enthusiastically reaffirmed. In the case of Democritus one finds that his doctrine was restored in the seventeenth century, after a lapse of some two thousand years; and it is interesting to note that both halves of the theory made their return at approximately the same time. Galileo—who differentiated between the so-called primary and secondary attributes and leaned towards atomism—was perhaps the first spokesman of the revival. And whereas Descartes propounded bifurcation but thought primarily in terms of continuous matter, we find that Newton already gave himself freely to chemical speculations of an atomistic kind. It is only that in those early days physicists lacked the means to quantify their atomistic speculations and put them to the test; not till the end of the nineteenth century, in fact, did 'atoms' begin to come within experimental range. But all along the atomistic conception of matter had played a decisive heuristic role; as Heisenberg points out, 'The strongest influence on the physics and chemistry of recent centuries has undoubtedly been exerted by the atomism of Democritus.'[6]

In the course of the twentieth century, however, the picture has begun to change. First of all a number of powerful and influential philosophers have at last appeared upon the scene—Husserl, Whitehead, and Nicolai Hartmann, for example—to challenge and refute the Cartesian premises; and meanwhile other types of philosophies have also come into vogue, such as pragmatism, neopositivism, and existentialism, which do not so much disqualify as bypass the bifurcationist axiom. Thus, whether by refutation or neglect, it can in any case be said that Cartesianism has now been abandoned by the leading philosophic schools.

In the scientific world, on the other hand, it is the Democritean doctrine of atomism that has found itself under attack, while the bifurcationist premise has remained virtually unquestioned. And even when it comes to atomism—which is plainly at odds with the latest findings of particle physics—it turns out that not a few leading physicists remain tacitly Democritean in their *Weltanschauung*; which is precisely why Heisenberg laments that 'Today in the physics of elementary particles, good physics is unconsciously being spoiled by bad philosophy.'[7] Few however realize that both halves of this 'bad philosophy' are with us still, and must be given up if one is to make philosophic sense out of present-day physics.

Meanwhile it is bifurcationism that poses the greater problem. In the first place, bifurcation is far more fundamental, and consequently far more difficult to comprehend; but most importantly, it happens to be the premise upon which the totalist conception of physics is based. Physicists can well do without atomism, but are in general loath to relinquish their totalist claims; and so they are committed, like it or not, to the Cartesian hypothesis.[8]

6. *Encounters with Einstein* (Princeton, NJ: Princeton Univ. Press, 1983), p 81.

7. Op. cit., p 82.

8. One consequently believes in bifurcation for much the same reason that one believes in Darwinian evolution: for indeed, so long as one insists that every phenomenon of Nature can in principle be understood by the methods of physics alone, both dogmas prove to be indispensable. My views on this question have been detailed in *Cosmos and Transcendence*, op. cit., chap. 4; *Theistic Evolution* (Tacoma,

❖   ❖   ❖

If the act of perception does put us in touch with the external world—as I claim—the question remains, of course, how this prodigy is accomplished. In the case of visual perception (to which we may as well restrict our consideration), there exists no doubt the perceptual image of an external object; and yet what we actually perceive is not the image as such but the object, precisely. We 'see' the image, if you like, but perceive the object; for in a sense we perceive more than we see, more than is given to us or passively received. And thus perception is not sensation, pure and simple, but sensation catalyzing an intelligent act.[9]

It should however be noted that the perceptual act is not rational or ratiocinative: there is absolutely no reasoning involved in perceiving an object. If the perceptual act were ratiocinative, moreover, it would be a matter of interpreting the image as representing an external object; and this would imply, firstly, that the object is conjectural—a concept as distinguished from a percept—and secondly, that the image, for its part, is viewed as image, which it is not. The point is this: In the perceptual act the image is viewed, not as image, but as a part or aspect of the object; it is seen, in other words, as something that belongs to the object, even as the face of a man belongs to the man. The image thus becomes more than an image, if one may put it thus: it is perceived as a surface, a face, an aspect of a thing which immeasurably transcends the image as such.

Now this decisive transition—from image to aspect—is something that reason or reasoning can neither effect nor indeed comprehend—which may well account for the fact that philosophers have experienced so much difficulty in coming to grips with the problem of perception. We have as a rule forgotten that there is

WA: Angelico Press/Sophia Perennis, 2013), chap. 1; and *Cosmos, Bios, Theos*, edited by Henry Margenau and Roy A. Varghese (Chicago: Open Court, 1992).

9. Thus we perceive the object as three-dimensional even though the image is plane. The conceivable objection that stereoscopic vision is due to the fact that there are two images is beside the point for two reasons: firstly, because we do not in reality see two images, but only one; and secondly, because even when we look at a familiar object with just one eye, we still perceive it as three-dimensional.

an intelligence which is intuitive, direct and instantaneous in its operation, an intelligence which has no need for dialectic or discursive thought, but flies straight to the mark like an arrow; and much less do we realize that this high and forgotten faculty—which the ancients termed 'intellect'—is operative and indeed plays the essential role in the act of sense perception. To discursive thought, image and object must remain forever separated—cut asunder, one could say—for it is the very nature of the ratiocinative faculty to analyze, to break apart. Thus, in the absence of intellect—if we were endowed, in other words, with no more than a capacity for the passive reception of images plus a faculty of reason—authentic perception would be impossible, which is to say that the external world would become for us a mere conception or speculative hypothesis. Like Descartes, we could never see it, never touch it, never hear its sound.

It is by force of intellect that the perceived object is joined to the percipient in the act of perception—assuming, of course, that it is a *bona fide* or valid perception; for as I have noted earlier, the perceptual act can indeed miscarry, as happens, for instance, in the case of an optical illusion or a hallucinatory perception. As the ancients would put it, the perceptual act can miscarry because it is not purely intellective, but only 'participates' in the intellect; but these are questions which do not particularly concern us at the moment. For the present it suffices to take note of the fact that there is a non-ratiocinative mode of intelligence by which the transition from perceptual image to perceived object is effected, and that reason or discursive thought is simply not equal to the task. But of course this does not in the least imply that there is anything irrational in the perceptual act, or better said, in the philosophical acknowledgement that we do actually look out upon the external world.

It might not be out of place to observe, in connection with what has just been said on the subject of human intelligence, that the reduction of intellect to reason—which is the fallacy of rationalism—might well constitute the prime offense, not just of René Descartes and his more immediate followers, but perhaps of modern philosophy in general. For even the anti-rationalist schools, such as pragmatism and existentialism, seem to presuppose the same reduction, the same rationalist denial of intellect. But be that as it may, once this philosophically fatal assumption has been made we find ourselves caught up in a dichotomy which cannot by any means be bridged. The external world of matter and the internal world of mind, if you will, have then seemingly lost their connection; and this means, of course, that the universe, and our position therein, have become *de facto* unintelligible. It is the nature of reason to analyze, to cut asunder even, it would seem, what God Himself has joined; no wonder, then, that a *Weltanschauung* based upon reason alone should turn out to be fractured beyond repair. Intellect, on the other hand, is the great connector; it unites what appears disparate, not externally, to be sure, but by bringing to light a deep and pre-existent bond. To put it in somewhat mythical terms, what 'all the king's horses and all the king's men' have failed to 'put back together again,' the 'royal intellect' restores in a trice.

Now the classic example of this marvelous feat is no doubt the ordinary, humble act of sense perception: the act, for instance, of beholding an apple. The chasm between subject and object—the epistemological abyss that had baffled a Descartes and a Kant—is bridged, I say, in the twinkling of an eye; every child can and indeed does pull off the miracle—which does nothing, however, to lessen its magnitude. For it is and remains a marvel—seeing that the apple is outside of us, and we perceive it nonetheless. Or in the words of Aristotle: that in the act of knowing 'the intellect and its object unite.'

Let no one, moreover, deny the miracle: that 'through' the image ('as through a glass') we perceive the object itself, the external thing. Let there be no mistake about it: the term of the intentional act is not simply another image, or a subjective representation, but the object itself; what we perceive is the apple, precisely, and not just a picture, or a concept, or an idea of the apple. But of course our perception or knowledge is incomplete: 'For now we see through a glass, darkly; . . . now I know in part' (1 Cor. 13:12).

It is no small thing that transpires thus in these familiar daily acts; for the intelligence manifested therein is mysterious: a power so awesome that its very existence within us belies our usual notions as to what man is, and how he came to be.

❖   ❖   ❖

Let us consider how one commonly envisions the perceptual act. An external stimulus impinges upon a sense organ (the retina, let us say) and causes a current of coded information to be conveyed along neuron paths to appropriate brain centers. But what happens then? A majority of scientists, perhaps, still espouse the old materialist or 'monist' position to the effect that the brain is everything, which is to say that the psychic life is viewed as an epiphenomenon of brain function. On the other hand, a growing number of neurophysiologists and brain experts—including some leading authorities—have come to believe that the monist position is untenable, and that the phenomena of perception and thought can only be explained on the assumption that, in addition to the brain, there exists also a 'second element' or mind. As a noted brain surgeon has put it:

> Because it seems to be certain that it will always be quite impossible to explain the mind on the basis of neuronal action within the brain, and because it seems to me that the mind develops and matures independently throughout an individual's life as though it were a continuing element, and because a computer (which the brain is) must be operated by an agency capable of independent understanding, I am forced to choose the proposition that our being is to be explained on the basis of two fundamental elements.[10]

One is sorely tempted to regard the second element or mind as a kind of ghost within the machine—presumably because one does not know how else to conceive of it. And this brings into play the unsettling notion of a conscious agent able to decipher the states of a billion neurons and integrate this information into a perceived image—all in a fraction of a second! But it is not in fact the speed of the operation or its complexity that baffles us, but its nature: for

---

10. Wilder Penfield, *The Mystery of the Mind* (Princeton, NJ: Princeton University Press, 1975); quoted by E.F. Schumacher in *A Guide for the Perplexed* (New York: Harper & Row, 1977), p76.

neither a mechanism nor a human observer could remotely accomplish such a task.

But let us suppose that somehow the mind is able to 'read the computer': to transform neuronal information into a perceptual image: what then? The resulting scenario of the perceptual act is evidently tantamount to an observer watching monitors hooked up to an external source. One might think that by now all is well, and that having come thus far one has at last arrived at a viable model. But such is not the case: for what our observer perceives is obviously a picture on a monitor and not the external object at all. Now, from an information-theoretic point of view this poses no problem, and there is in fact no significant difference between supposing that the observer does or does not perceive the outside world; for example, if it be a question of reading an external instrument, it is evidently immaterial whether he looks at the monitor or directly at the external scale. But then, what we are seeking to understand is not just the transmission of information (in the sense of the electrical engineer), but the phenomenon of perception, which is something else entirely—even though it does obviously entail a transmission of the stipulated kind. We must remember that authentic perception terminates, as we have seen, not in a mere image, but in a face or aspect of the external thing itself. But here the observer/monitor model fails: there is no getting around the fact that what our observer perceives is the monitor, and the monitor alone. In short, the given model as it stands turns out to be incurably bifurcationist. It may do justice to the brain, but fails to comprehend the second element: the mind and its powers.

There is an ancient and long-forgotten belief to the effect that the perceiving eye sends forth a 'ray' to meet the object; and whereas this notion may strike many nowadays as just another 'primitive superstition,' is it not conceivable that the afferent propagation from object to percipient needs indeed to be complemented by an efferent process, a propagation going the other way? And if science has found no trace of such an efferent 'ray,' could this not be due to the fact that its methods are unsuited for the detection of that process? Thus, if the afferent propagation be 'material,' might not the efferent be, let us say, of a 'mental' kind? It appears to me that when it comes to the problem of perception[11] we are scarcely in a position to reject 'strange' doctrines out of hand. All that we know, at this point, is that the pieces presently within scientific range do not fit together—which seems to imply that the missing piece of the puzzle *must* in fact be 'strange'. Call it 'mind', 'spirit', or what you will; as Sir Charles Sherrington has observed: 'It goes in our spatial world more ghostly than a ghost. Invisible, intangible, it is a thing not even in outline, it is not a thing.'[12] One cannot but agree with the eminent neurophysiologist that science 'stands powerless to deal with or describe' that elusive and enigmatic presence, by which apparently the perceptual act is consummated.

❖   ❖   ❖

By the 'corporeal world' we shall henceforth understand the sum total of things and events that can be directly perceived by a normal human being through the exercise of his sight, his hearing, and his senses of touch, taste and smell; which is to say, in a word, that the corporeal domain is no more and no less than the actual world in which we normally find ourselves. But of course this affirmation, simple and indeed obvious as it is, will immediately be challenged by the bifurcationist, on the grounds that what we actually perceive is not a world at all—not an external reality—but a private phantasm, of which only certain quantitative features have an objective significance. In other words, what on a pre-philosophic level we take to be the world is thus denied external or objective status—to make way, presumably, for the world as conceived by the physicist. The recognition, therefore, of what may be termed the principle of non-bifurcation amounts to a rediscovery—or a reaffirmation, if you will—of the corporeal world, a world which according to Descartes and his disciples does not exist.

In reality, of course, it is plain that no one has paid the slightest attention to the Cartesian authorities; which is to say that in our

---

11. Cf. *Science & Myth* (Philos-Sophia Initiative, 2023), chap. 4.
12. *Man on His Nature* (Cambridge: Cambridge University Press, 1951), p 256.

daily lives we do not question, let alone deny the authenticity of the sense-perceived world. Everyone continues to go about his business, firm in the conviction that 'Mountains are mountains and clouds are clouds' as the Zen master points out; and yet we do, most of us, have our Cartesian moments. Try, for instance, to persuade a university professor, or even better a graduate student, of non-bifurcation, and you will soon bring out the Cartesian in him; such is the force of education. But such, too, is the nature of the question; for indeed, what is obvious in the unreflective state is not *ipso facto* true—as if thoughtlessness alone could bestow infallibility. Cartesian doubts, therefore, are far from illegitimate, and what we take exception to are not in fact the doubts, but the philosophy.

Yet this philosophy has been very much bred into us through the educational process, so much so that it may come as something of a shock to be told outright that the perceived world is indeed real, and that we are *not* after all mistaken during most of our waking life— throughout the hours and days during which we remain unheedful of the bifurcationist teaching. To be sure, this conspicuous resistance and disbelief with which most of us react to the principle of non-bifurcation when it is asserted may indeed seem strange, given the fact that at all other times, both before and after the philosophic interlude, we remain staunchly committed to the principle in question. It is only when non-bifurcation is explicitly affirmed that we generally turn against it and blithely deny what otherwise we firmly believe. In short, the Cartesian philosophy has plunged us into a collective state of schizophrenia, a doubtless unwholesome condition, which may well have something to do with not a few of our contemporary ills.

But be that as it may, it is no easy task to cut the Newtonian knot and cast off the burden of an antinomous philosophy; for even though bifurcation as such may hold no particular attraction, it does confer the considerable benefit of apparently bolstering the claims of a physics that would be totalist in its scope. Add to this the widespread belief to the effect that the prevailing *Weltanschauung* has been mandated by the positive findings of an exact and unerring science—and one begins to discern the magnitude of the problem. It is no wonder then that the philosophic foundations of physics are in disarray. More than half a century has now elapsed since Whitehead first lamented this state of affairs and lectured us on what he termed 'a complete muddle in scientific thought, in philosophic cosmology, and in epistemology';[13] but the muddle remains, and if anything, has only been exacerbated by a rash of pseudo-philosophical writings which do little more than pour new wine into the old bottles. So far as the physicists are concerned, moreover, it would seem that most are little inclined to the investigation of philosophic foundations; and even then it appears that their scientific prowess does not always carry over into the philosophic domain. As Heisenberg has well said:

If one follows the great difficulty which even eminent scientists like Einstein have in understanding and accepting the Cophenhagen interpretation of quantum theory, one can trace the roots of this difficulty to the Cartesian partition. This partition has penetrated deeply into the human mind during the three centuries following Descartes, and it will take a long time for it to be replaced by a really different attitude toward the problem of reality.[14]

13. *Nature and Life* (New York: Greenwood, 1968), p 6.
14. *Physics and Philosophy* (New York: Harper & Row, 1958), p 81.

II

WHAT IS THE
PHYSICAL UNIVERSE?

ONE WOULD LIKE TO SAY that the physical universe is simply the world as conceived by the physicist; but then, it is far from clear just how the physicist does conceive of the world. One must remember, in the first place, that physics has undergone a stupendous development, and continues to progress by leaps and bounds. And what is more, there has of late been little agreement among physicists as to what it is, exactly, that physics is bringing to light. How, then, can one speak of the 'the world as conceived by the physicist'?

One can do so, up to a point, by virtue of the fact that physics has a methodology of its own, a distinctive mode of inquiry. Particular physical theories may be superseded, and philosophical opinions may come and go; but the basic cognitive means by which physics as such is defined remain unchanged. And these cognitive means determine their objects in a general way: that is the crucial point. Let us say, then, that the physical universe is the realm of things knowable in principle by these particular means—and consider whither this leads.

We have seen in the preceding chapter that the corporeal world exists 'for us': as the domain of things to be known by way of sense perception; and now we find that the physical universe exists 'for us' in much the same sense. It is only that the respective means of knowing are markedly different. In the first case we know through direct perception, and in the second through a complex *modus operandi* based upon measurement—which is something else entirely.

Let us briefly examine the act of mensuration. The first thing to be noted is that one measures, not directly by sight, or by any other sense, but by means of an artefact, an appropriate instrument. What counts, in fact, is the interaction between object and instrument: it is this that determines the final state of the instrument, and hence the outcome of the measurement. And that outcome, moreover, will be a quantity; a number, if you like. Now to be sure, the experimental physicist makes use of his senses at every step; and it is by way of sense perception, in particular, that he ascertains the final state of the instrument. But this does not mean that he perceives the quantity in question. Let us be clear about it. Strictly speaking, one does not perceive such a thing as the weight or diameter of a familiar object, any more than one is able to perceive the magnetic moment, say, of the electron. What one perceives are corporeal objects of various kinds—including scientific instruments. And of course one is able to read the position of a pointer on a scale. But one does not perceive measurable quantities. And that is the reason why one needs an instrument. The instrument is required precisely because the quantity in question is *not* perceptible. It is thus the function of the instrument to convert the latter, so to speak, into the perceptible state of a corporeal object, so that, by means of sense perception, one may attain to the knowledge of something that is not in itself perceivable.

Now the *modus operandi* of physics is based upon measurement, as I have said; it is thus through acts of measurement that the physical universe comes into view. The physicist looks at reality—not with the ordinary human faculties of perception—but through artificial instruments; and what he sees through these man-made 'eyes' is a strange new world comprised of quantities and mathematical structure. In a word, he beholds the physical universe, as distinguished from the familiar corporeal world.

What, then, are we to make of this curious duality? Can we say, for example, that one of the two domains is real, and the other subjective, or somehow fictitious? It appears that there are actually no cogent grounds in support of either reductionism. What you see depends on the 'lenses' through which you look: that is the gist of the matter.

The question arises how the two apparent worlds—or 'cross sections of reality'—can coexist, or fit together, as indeed they must.

And suffice it to say, for the present, that this is a matter which cannot be investigated, or understood, through the cognitive means associated with either realm. Neither through sense perception nor by the methods of physics can the issue be resolved—for the simple reason that each of these cognitive means is restricted to its own sphere. What is needed, ideally, is an integral ontology, and we may as well leave unresolved, for the time being, the question whether such an enterprise is feasible. What matters, meanwhile, is the realization that each of our two domains—the physical no less than the corporeal—is limited in its scope. In each case there are the things that can be known by way of the given cognitive means, and there are things that can not. Like a circle, the concept of either domain includes and excludes at one and the same stroke. And there should be no doubt from the start that what each excludes must, in fact, be immeasurably more vast than the multitude—staggering though it be—of its own total content.

❖   ❖   ❖

The physical universe 'comes into view' by way of measurement; but one should immediately add that measurement alone does not suffice. There is also perforce a theoretical side to the cognitive process, which is to say that nothing pertaining to the physical domain can be known without a theory, an appropriate 'model.' The experimental and theoretical aspects of the discipline, moreover, work hand in hand; there is a marvelous symbiosis between the two—too delicate, perhaps, to be adequately described in textbook terms. Suffice it to say that experiment and theory combine to make a single cognitive enterprise, a single 'way of knowing'.

Physical objects, then, are to be known by means of a suitable model, a theoretical representation of some kind. To be sure, object and representation do not coincide. We know the object *through* the representation—even as one knows a corporeal object by way of a mental image. The theoretical representation serves thus as a symbol, a sign—which is however indispensable. For indeed, one cannot know or even conceive of a physical object except by way of a model, a theoretical construct of one sort or another. Certainly we can and often do change the representation of a given physical entity; but we cannot let go of one model without the aid of another—on pain of losing the object entirely.[1]

Let us note (for the sake of maximal clarity) that if the object were indeed reducible to the representation, it would obviously not be subject to measurement; a mere model, after all, does not affect our instruments. Physical objects, on the other hand, do evidently affect the appropriate instruments of measurement—by definition, if you like; and this means that they have a certain existence of their own. The passage from representation to object, therefore, constitutes an intentional act no less enigmatic, certainly, than the humble act of sense perception, considered at length in the preceding chapter. It is no wonder, then, that this crucial step pertaining to the *modus operandi* of physics has not been well understood. That the intentional act of the physicist, so far from being accessible to mankind at large, presupposes evidently an exacting apprenticeship—not to mention certain specific intellectual qualifications which perhaps not everyone may possess in full measure—all this does not conduce to facile understanding of the epistemological issue. But most importantly, it happens that the philosophic premises to which we are nowadays prone do not in reality allow any knowledge of physical objects— any more than they allow the perception of corporeal entities. Meanwhile every reputable physicist has learned to consummate the intentional act of which we speak, and does no doubt consummate that crucial act repeatedly in the exercise of his professional labors— even though, *qua* philosopher, he may well be committed to a school of thought which denies the very possibility of this cognitive act. The scenario is altogether reminiscent of the bifurcationist who denies—once again, in his philosophic moments—the possibility of sense perception: the quotidian act by which we look out upon the world and perceive—not mere sense data, or mental representations—but a myriad existent things. We have already commented sufficiently (in chap. 1) on this strange phenomenon, which now we

---

1. I shall use the term 'physical system' to denote a physical object as conceived in terms of a given theoretical representation. Different representations of one and the same physical object give rise thus to different physical systems.

encounter once more in the context of scientific knowing. The point, in either case, is that it is one thing to know, and quite another to know *how* we know.

The model through which a physical object is known must of course square with the measurable facts; which is to say that it must be possible to derive empirically verifiable consequences therefrom. The representation has thus a certain operational meaning or empirical content, failing which it would have no connection with the scientific enterprise. We need, however, to understand clearly that it has, in addition, an intentional content, which is to say that it serves as a sign or symbol, whose referent is the physical object itself. The two kinds of content or meaning—the intentional and the operational—are moreover closely linked: for indeed, a physical object can be modeled or represented precisely by virtue of the manner in which it presents itself to empirical observation. But even though we may know the object by the way it affects our instruments, we conceive of it nonetheless as the external or transcendent cause of these observable effects, and not simply as their sum-total. An electromagnetic field, for example, is assuredly more than a set of instrument readings, and a proton more than an ensemble of tracks in a cloud chamber. Contrary to what the positivists would have us believe, the physical object cannot in fact be reduced to its observable effects. The object, thus, is not the manifestation, but the entity, rather, which manifests itself. Our instrument readings and condensation tracks point therefore beyond themselves; and that is precisely the reason, of course, why these readings and sightings are of concern to the physicist. His primary interest is not in positivistic games, but in a hidden reality which manifests itself, at least partially, in all kinds of measurable effects. The physical universe, no less than the corporeal, is thus in a way transcendent—even though (as I have said repeatedly) it exists 'for us'.

❖   ❖   ❖

Strictly speaking, no one has ever perceived a physical object, and no one ever will. The entities that answer to the *modus operandi* of physics are by their nature invisible, intangible, inaudible, devoid of taste and scent. These imperceptible objects are conceived by way of mathematical models and observed by means of appropriate instruments. There are, however, physical entities which present themselves, so to speak, in the form of corporeal objects. Or to put it the other way round: Every corporeal object X can itself be subjected to all kinds of measurements, and determines thus an associated physical object SX. If X is a billiard ball, for instance, we can measure its mass, its radius, and other physical parameters, and can represent the associated physical object SX in various ways: for example, as a rigid sphere of constant density. The crucial point, in any case, is that X and SX are not the same thing. The two are in fact as different as night and day: for it happens that X is perceptible, while SX is not.

Now the first of these claims is obvious and incontrovertible. Everyone knows that such a thing as a billiard ball is perceptible. Or better said, everyone knows this full well—so long as he is not a bifurcationist. But what about SX: why is this *not* perceptible? There are those, presumably, who would argue that a rigid sphere, for instance, can very well be perceived. But whereas, strictly speaking, this turns out not to be the case,[2] the contention is actually beside the point. For the question before us is not whether such things as rigid spheres can be perceived, but whether SX can be, and that is quite another matter. For whereas the associated physical object SX of the present example can indeed be represented (within certain limits of accuracy) as a rigid sphere, it can also be represented in many other ways. For instance, as an elastic sphere—a model which in fact can give rise to a more accurate description. More importantly, however, one knows today that physical objects are composed of atoms—or more generally, of subatomic particles—and that all continuous or 'classical' representations convey no more than a

---

2. At the risk of flogging a dead horse, the point could be argued as follows: A rigid sphere of constant density is entirely characterized by two numerical constants: its radius R and density δ. Neither R nor δ, however, can be perceived (these quantities can of course be measured, but as noted before, to measure is not the same as to perceive). But since the quantities in terms of which the rigid sphere is defined are imperceptible, so is the rigid sphere. Or again: No one has ever perceived (in the visual sense) an object bereft of all color. But the rigid sphere has no color (it is characterized by R and δ, as I have said). Hence it is imperceptible.

rough and partial view of the entity in question. But now, if we suppose that SX is indeed an ensemble of atoms, or of subatomic particles, is it yet conceivable that SX can be perceived? Obviously not; for it is plain that what we perceive is not a collection of atoms, subatomic particles or Schrödinger waves, but a billiard ball, precisely. It could of course be claimed that the ensemble of atoms or particles gives rise somehow to the perceived or perceptible object—but this is an entirely different question. What concerns us at present is the identity of this perceived or perceptible object, and not its conjectured cause. And this identity is beyond dispute: what we perceive is the red or green billiard ball, to say it once more. No one, to repeat, has ever perceived an ensemble of subatomic particles, or a collection of atoms.

One arrives thus at a basic recognition which has long been obscured on account of the bifurcationist bias: one finds now that every corporeal object X determines an associated physical object SX. We shall henceforth refer to X as the *presentation* of SX. Not every physical object, of course, has a presentation; which is to say that we can distinguish between two types or classes of physical entities: those which do and those which do not admit a presentation. *Subcorporeal* and *transcorporeal*, let us say. But I hasten to point out that this dichotomy has to do, not with the physical objects as such, but with their relation to the corporeal domain. The physicist, in other words, who investigates the structure or physical properties of the objects in question will find no trace of this dichotomy. As atoms congregate into molecules, and molecules become joined into macroscopic aggregates, there is no point, no magic line of demarcation, signaling the start of the subcorporeal realm. For indeed, it is only with reference to the corporeal plane that this notion is defined. And, therefore, if we had eyes only for the physical plane—and could see only atoms and the like—there would be no way by which we could distinguish subcorporeal from transcorporeal aggregates.

The distinction, however, is nonetheless vital to the economy of physics. For it is plain from what was said earlier that instruments of measurement must be corporeal. The process of measurement must terminate, after all, in the perceptible state of a corporeal object. But this means, in light of the preceding considerations, that the *physical* instrument is perforce subcorporeal; to be precise, it must be the SI of a corporeal instrument I.

※   ※   ※

It is to be noted that, in addition to measurement, physics has need of empirical procedures which terminate, not in the numerical value of a physical parameter, but in a pictorial representation of some kind. Examples are legion, and range from the various forms of telescopy to electron microscopes and bubble chambers. Now, in all these cases a physical object or process is somehow converted into a pictorial display—a photograph, for instance—which incorporates information regarding the object or process in question. And this information, moreover, is once again quantitative, but not specifically numerical, if one may put it so. One is consequently obliged to distinguish between measurement in the strict sense—which terminates in a numerical value, or in a 'pointer reading'—and a second mode of scientific observation, which (for lack of a better term) we shall refer to as 'display.' The two modes can of course be combined, as happens, for instance, when a photograph—the terminus of a display—is used to effect measurement. But the symbiosis can also proceed in the opposite direction—as in the case of graphic displays incorporating the results of a set of measurements. Despite the close interconnection of the two modes, however, I contend that neither can be assimilated or reduced to the other without violence, which is to say that physics has need of both.[3]

In view of the fact that physical objects are not perceptible, the question arises now in what sense one can speak of a 'pictorial representation' of something which in principle cannot be perceived.

3. One might mention that display and measurement correspond precisely to the two so-called modes of quantity: extension and number, which until modern times were conceived to be irreducible. It was Descartes who blurred the distinction through the invention of what has come to be known as analytic geometry. But be that as it may, the distinction between extension and number persists, and despite the fact that just about everything these days can be 'digitalized', the need for pictorial representations is with us still.

The temptation is great to imagine that the display conveys a likeness—much as an ordinary photograph conveys a likeness of the corporeal object which it depicts. But how can one speak of a likeness if one has never seen the original, and if indeed the original can not be seen at all? To tell whether a portrait, for example, is a likeness or not, one needs, after all, to look at the subject; but if the subject cannot, in principle, be looked upon, then it has no 'look' at all, and it no longer makes sense to speak of a likeness.

Such is the case when it comes to display. Likeness in the usual sense is out of the question. But if ordinary likeness fails, there must yet be a similarity of some kind, in the absence of which it would make no sense to speak of display. There is, consequently, a notion of likeness applicable to display, and it should come as no surprise that the similarity in question is a likeness of mathematical form, of abstract structure. The fact that mathematical forms can on occasion be depicted in visual terms is of course familiar to every student of mathematics; anyone who has taken a course in calculus or analytic geometry, for instance, will recall the parabolic graph of the function given by the formula $y = x^2$. And one will also understand full well that the function as such is imperceptible, and that the graph is not simply a likeness in the usual sense. At the same time, however, one understands that the graph does in a sense depict the function; after all, one can see from the graph that it assumes a minimum at $x = 0$, that the derivative vanishes at that point, that the absolute value of the derivative increases with the absolute value of x, and so forth. Moreover, one can in principle recover the function from its graph; which is to say that if we neglect the fact that ordinates and abscissas cannot in practice be ascertained with arbitrary precision, one is able to obtain from the graph the value $f(x)$ of f for every x.

In the case of scientific display, of course, the object is not a mathematical but a physical entity of some kind; but yet the physical entity owns mathematical form, and it is this form, precisely, that is being depicted. The case of display is thus indeed analogous to that of the graph. For it goes without saying that if the physical entity were in addition to have properties of a non-mathematical kind, these would not show up under display. In other words, what the photograph, let us say, can have in common with the radio star, or with an ensemble of interacting particles—or for that matter, with our erstwhile rigid sphere—can be nothing more nor less than a mathematical form.

Any number of examples can be given to illustrate this basic point. Consider, for instance, an X-ray photograph of an ordinary solid object. If we coordinatize the region in a Cartesian way and suppose that the X-rays travel parallel to the z-axis, and that the photographic plate is set parallel to the xy-plane, then the photograph itself can be coordinatized by x and y, and the optical density $f(x,y)$ of the emulsion (after exposure and developing) will define a certain function f. Now, it is this function f, precisely, that the photograph shares, so to speak, with the body in question; for indeed, if we knew the appropriate 'optical density' $\delta(x,y,z)$ of the object, integration with respect to z would give an 'effective optical density' $\delta^*(x,y)$, and knowing δ^* one could calculate f. The utility of X-rays, on the other hand, derives from the fact that this calculation can be inverted: knowing f one can obtain δ^*. The purpose of X-ray display, one might say, is to exhibit the function δ^*. It is δ^* that we discern when we examine an X-ray transparency; one sees where δ^* is small or large by the comparative blackness or whiteness of the region, and can judge the steepness of its increase or decrease in various directions.[4] One can in fact regard the transparency as a 'graph' of a function of two variables, in which function values are depicted by a 'density of dots'.

There are, of course, many other types of display; and it is to be noted, in particular, that there is no reason why the coordinates x and y of a display should invariably represent spatial dimensions. The familiar example of the oscilloscope is instructive in that regard. In its simplest mode of operation there is just one input $V(t)$, where V is a voltage and t represents time. The monitor will then exhibit a graph of the function V in which the ordinate represents voltage

---

4. For medical purposes one is of course interested in ascertaining, not just $\delta^*(x,y)$, but $\delta(x,y,z)$; it is δ that is needed, after all, to pinpoint a tumor, or a minute foreign object. Let us add that this is the subject of a mathematical discipline known as tomography, which underlies the technology of 'scans'.

(and thus, too, whatever the input voltage itself may represent) and the abscissa represents time. One can thus 'see' an electric pulse, a sound wave, a temperature fluctuation, or whatever it may be that has been represented by an input V(t). Or again, one can use the oscilloscope with two inputs—V(t) and W(t), let us say—and let the monitor exhibit the relation between V and W as given by a curve in the VW-plane. In all instances of display, however, what is being exhibited is either a function or a more general relation; the physical system as such, on the other hand, remains invisible.

❖   ❖   ❖

It is to be noted that every form of scientific observation—whether it be a case of measurement or of display—hinges upon the correspondence between a corporeal object X and the associated physical object SX. It hinges, in other words, upon an act of presentation (X being the presentation of SX). In general, the transition from the physical to the corporeal domain, which consummates the process of observation, is to be effected precisely by a passage from SX to X; for indeed, we know of no other link or nexus between the physical and the corporeal levels of existence. Moreover, it is evident that the experimental physicist avails himself of this connection constantly, and as a matter of course. He does so, for instance, when he treats a corporeal object as a physical system, or when he employs corporeal entities to 'prepare' a physical system of a transcorporeal kind; and he does so, to be sure, when he measures or displays a physical object.

It happens, however, that this crucial link is nowhere recognized. Thus, first of all, it does not show up on the physicist's maps, for the simple reason that these maps refer exclusively to the physical domain (and are bound, therefore, to exclude the link in question). Nor is there any room for it in our customary scientistic world picture; for this Cartesian or 'classical' *Weltanschauung*, as one knows, is based upon the bifurcation postulate. It consequently negates the existence of the corporeal domain, and thus, too, the existence of a link. Yet, recognized or not, the link of presentation is there, and seems in fact to be in constant scientific use. The circumstance that we do not understand this nexus—whether by means of physics or of philosophic inquiry—seems not to matter in the least. Does one not also make ample use of sense perception—which proves to be no less incomprehensible?

It comes down to this: There can be no knowledge of the physical domain without presentation—even as there can be no knowledge of the corporeal world in the absence of sense perception. There is no way, of course, to convince the die-hard skeptic that the physical universe exists in the first place, let alone that it can be known; and it is always possible, certainly, to relapse into a positivistic reductionism. Suffice it to say, however, that one cannot avoid the idea of presentation—except at the cost of the physical universe.

The question arises now: What can we learn about a physical object from its presentation? Despite the fact that X and SX are as different as could be—think of a red billiard ball, for instance, and of a cloud of atoms—there must yet be a certain 'resemblance' between the two, failing which X could tell us nothing about SX; what, then, is that 'resemblance' or connection? Now the first thing to be noted in that regard is that X and SX occupy exactly the same region of space—strange as this may seem.[5] For indeed, it would make no sense at all to distinguish between a so-called corporeal and a physical space—the reason being that the physical space would have no meaning unless we could relate it to the corporeal, which can only be done, however, by way of presentation. But this would be tantamount to an identification of the two spaces, and thus to the spatial coincidence of X and SX.

But this spatial coincidence implies that the notions of distance and angle—which can be defined, as one knows, in terms of operations involving corporeal measuring rods—carry over to the subcorporeal domain. Every decomposition, therefore, of a corporeal

---

5. The fact that X and SX occupy the same region in space is not in the least paradoxical. First of all, it does not contradict our sensory experience, because perception pertains only to X. From a theoretical point of view, moreover, there is nothing contradictory in the notion of two entities occupying the same space; this happens, for instance, in the case of fields. An electric field, for example, can coexist with a magnetic, or gravitational field. Once again, what you see depends on how you look.

object X into corporeal parts, corresponds to a congruent or geo-metrically isomorphic decomposition of SX. In a word, there is a 'geometric continuity' between X and SX.[6] And it is by virtue of this geometric continuity, precisely, that physical objects can be observed. Thanks to this continuity one is able, for instance, to ascertain the state of a physical instrument from the position of a pointer on a scale (a corporeal pointer, needless to say, on a corpo-real scale). Or to put it in more general terms: the state of a physical instrument, as given by its internal geometry—or more exactly, by the relative positions of its subcorporeal parts—is passed on to the corporeal plane via presentation. All measurement, clearly, and every conceivable form of display, depend upon this fact.

One further remark: By virtue of geometric continuity, presenta-tion constitutes a mode of display. It constitutes indeed what could be termed the primary mode of observation, the point being that all other forms of observation are dependent upon presentational dis-play, as we have noted before.

❖   ❖   ❖

Whether it be a case of measurement or display, one observes a physical object in a scientific sense by causing that object to interact with a subcorporeal instrument; the effect or outcome of that inter-action is then transmitted to the corporeal level by way of presenta-tion. It must not be thought, however, that to observe the object in question one has simply to set up the appropriate equipment, wait for the desired interaction to occur, and take note of the result. For indeed, that result can be nothing more than a pointer reading, a numerical printout, or a graphic display of some kind. What the instrument yields, in other words, is data; but this is not what the physicist is after. Data is a means, to be sure, but not the end of the

6. There is also, of course, a 'temporal continuity' between X and SX. This means, first of all, that a corporeal object X, considered at a particular instant of time, constitutes a presentation of SX *at the same instant*; and secondly, that the notion of 'temporal distance' or duration, as measured by corporeal clocks, carries over to the subcorporeal realm.

observation process. What the physicist seeks, obviously, is the physical object; or better said: a certain knowledge or intellectual apprehension of the object. And this no scientific instrument—no empirical *modus operandi*—can give.

Observation, therefore, is not effected by empirical means alone. There can be no *bona fide* observation without the theoretical aspect of the enterprise coming into play. One might put it this way: To observe in the sense of the physicist is to pass from the perceptible to the imperceptible—and only theory can span the gap. As we have pointed out before, theory and experiment work hand in hand. The two combine to constitute a single cognitive act, a single 'way of knowing'.

Strictly speaking, there is no such thing as an 'empirical fact'—so long, of course, as the term is understood to exclude the concomitant role of theory. The circumstance, however, that nothing in the physical domain can be measured or displayed without the aid of theoretical premises serves in reality, not to cast doubt upon the validity of empirical findings, but to render theory itself more certain and indeed more 'evident' than it is commonly held to be. The customary notion that theoretical tenets are mere 'hypotheses' until they have been verified by experiment is thus overdrawn and somewhat misleading, for it happens that the supposedly 'solid facts of observation' can in principle have no more certainty than the so-called hypotheses upon which they rest.

What those who speak glibly of 'mere hypotheses' seem not to appreciate is the fact that intellect plays a rightful and most necessary role in the scientific process. Not just reason, or a capacity for logical thought, but intellect, in the ancient and traditional sense of a faculty for unmediated vision, whose objects are 'intelligible forms.' One has every right to suppose, moreover, that the great physicists are not only well endowed in that regard but know well enough how to make use of this high faculty in the course of their inquiries. In the best of cases, therefore, the premises laid down by the founders may indeed possess a kind of *a priori* validity which our textbook wisdom deems to be impossible.

One is reminded of an incident in the life of Albert Einstein which speaks eloquently to the point at issue. The year was 1919, and the Astronomer Royal of England had just announced, at a packed meeting of the Royal Society, that photographic plates exposed at the famous eclipse had confirmed the predicted bending of light. A telegram had been dispatched to Berlin, and someone forthwith burst into Einstein's office to break the news; but the great scientist seemed totally unmoved. 'What would you have thought if your theory had been disproved?' asked the young woman; 'Then I would feel sorry for the Old One,' was the reply.

The great fact is that the physical universe is not after all an unmitigated contingency. Contrary to what nominalists have been preaching for centuries, it is 'the universal in the particular' that bestows upon the particular its measure of being, which moreover coincides with its 'intelligible aspect'. And this implies that physics is in reality concerned, not with particular existents as such, but with particulars insofar as these exhibit a universal principle or law. Whatever may be left over remains of necessity unknown. Thus, what physics seeks, and is able to grasp in its own fashion, is the necessary in the contingent, or the eternal in the ephemeral, as one can also say.

One starts out, if you will, with the contingent in the form of empirical data. The data ensemble, however, is of interest precisely because it somehow mirrors or embodies a universal principle: it is this that the model or representation seeks to capture, as it were. But though the principle is in a way exemplified by the data, it is by no means disclosed, determined or forced upon us thereby. The representation constitutes thus 'a free creation of the human spirit,'[7] to put it in Einstein's words—which does not mean, of course, that it is merely subjective or altogether arbitrary. For indeed, what the representation depicts in its own way is an objective principle exemplified in the data, as we have said: the very same principle, in fact, that is first of all exemplified in the physical object itself. One and the same principle, thus, is reflected on three different levels: in the physical object, in the data ensemble, and in the model or representation. And that is the reason, after all, why the physical object is

---

7. A. Einstein and L. Infeld, *The Evolution of Physics* (New York: Simon and Schuster, 1954), p33.

knowable. Briefly stated, we know the object by way of the principle, and the principle by way of the representation, which in turn is
arrived at by way of the data ensemble.

It needs however to be understood that the passage from data
ensemble to representation is not to be effected by reasoning alone.
One does not arrive at Einstein's 'free creations' simply by logic, or
by following a set of rules; it is not a task that could be accomplished by a computer. And strictly speaking, the very apprehension
of the model or representation entails a certain intellective sight,
and thus involves the intellect, in the *bona fide* sense. The intellective act, moreover, by which we 'perceive' the representation, provides at the same time a certain apprehension of the principle itself.
In a sense, therefore, the physicist does 'see' the physical objects
with which he deals: he 'sees' them by way of their representation,
and thus in their principle or 'intelligible aspect'.

But that is just what those who speak glibly of 'mere hypotheses'
have failed to grasp: for indeed, where 'seeing' is concerned, there is
the possibility of 'seeing true', and of certitude as well. In a sense,
'seeing is believing' after all. And is this not the reason, finally, why
Einstein could remain unconcerned? Had he not seen the principle?
We surmise that this was indeed the case—which both explains and
justifies Einstein's laconic reply ('Then I would feel sorry for the Old
One').

The objection could be raised that inasmuch as physical theories
are perforce approximations, it therefore cannot be supposed that
they provide any true knowledge of physical objects, or could have
been arrived at by means of a quasi-infallible intellective act. But
why not? One must remember, first of all, that the physical universe
presents itself on various levels corresponding to the nature and
accuracy of the instruments through which it is observed. There is
nothing incongruous in the supposition that each level exhibits its
own laws or 'mathematical forms'—provided, of course, that the
laws associated with one level do not contradict those of another. In
particular, if level A happens to be more fundamental or 'accurate'
than level B, then the known laws pertaining to B should follow
from the laws pertaining to A—as seems in fact always to be the
case. Newtonian mechanics, for instance, can be derived from the

relativistic by assuming that velocities are small compared to the speed of light; or the thermodynamics of gases can be obtained *à la* Boltzmann from the classical mechanics of particles, which in turn can be conceived as a limiting case of quantum mechanics; and so forth. To be sure, from the standpoint of level A the laws pertaining to level B are approximate; but this does not in the least imply that the mathematical forms in question are 'merely subjective'—any more than the fact that a wheel, let us say, is not a perfect circle implies that its circular form or 'circularity' is therefore fictitious. If it be the case, in other words, that mathematical forms are not existentiated in the physical domain with 'absolute fidelity', it does not follow by any means that they are not existentiated therein at all. In short, one finds that each major theory does apply within its own proper domain, and that the founders did after all 'see true'. Where they may have erred, on the other hand, was in supposing that the laws in question apply without restriction. Newton, for example, did not foresee Einstein, and Einstein himself, as we know, experienced great difficulty in admitting the quantum realm. Each truly seminal physicist, perhaps, has a tendency to stretch his vision beyond its rightful bounds.

If the physical universe did not somehow embody or reflect mathematical forms, it would be simply unintelligible, and physics would not exist. Hence it does so embody or reflect mathematical forms, and in fact is constituted by these very forms: by its 'mathematical structure', precisely.

❖   ❖   ❖

What physics deals with, in the final count, are existentiated mathematical structures. It must however be admitted that layman and expert alike tend invariably to clothe these mathematical entities in more or less concrete imaginative forms derived no doubt from sensory experience. Or better said, we need in fact to clothe these intangible entities in sensory images of one sort or another to bring them within range, so to speak, of our mental faculties. In the case of the mathematician or informed physicist, moreover, this procedure is perfectly sound and plays indeed a vital role in the comprehension of structures and relations of a mathematical kind. In the hands of the expert the concrete form becomes a symbol—a catalyst of intellection, if you will. The accomplished theoretician knows full well how to extract from the concrete image an abstract form which may bear an analogy to the mathematical structure he wishes to comprehend. He has learned to seize what is essential and discard the rest. This is in fact the 'hidden art' that needs to be mastered. Following upon a more or less extensive apprenticeship one becomes at last proficient in the mental use of what might broadly be termed 'visual aids', which may range from simple images of material entities to such things as graphs and diagrams, not forgetting that even a mathematical formula bears necessarily a visual and syntactical aspect which also has its role to play.[8] It can thus be said of mathematics and physics no less than of any other human enterprise that 'Now we see through a glass, darkly'; generally speaking, sensible forms serve as a 'glass'.

The use of images or sensible supports, however, can easily become illegitimate and turn into a kind of intellectual idolatry. It all depends on whether we understand the difference between a visual representation—what the Scholastics would call a 'phantasm'—and the physical or mathematical object which it is supposed somehow to represent. The moment image and object are confused, error ensues; when the phantasmata are mistaken for the reality, fantasy ensues. But to tell the truth, the line is easily crossed and re-crossed, so much so that it may be more realistic to speak, not of pure knowledge versus unmitigated fantasy, but of degrees. The logical distinction, however, between a 'symbolist' and a 'concrete' employment of phantasmata retains its full validity and its rights, human weakness notwithstanding.

There are, then, degrees of comprehension, and even physicists are by no means exempt from the concretizing tendency. They too, in other words, are prone on occasion to 'reify' the physical object

8. It could be pointed out in this connection that language—and thus thought—has obviously its sensory support, auditory though it be. However, when it comes to the comprehension of mathematical structure it is doubtless visual symbols that play the primary role.

(as we shall henceforth say) through a more or less naïve acceptance
of visual supports; and it could even be argued that as a rule they
reify thus quite freely so long as the phantasmata in question do not
conflict too blatantly with the logical or mathematical demands of
their theory. And yet reification of even the most seemingly innocu-
ous kind is always illegitimate; in contrast to a genuinely symbolist
use of visual supports it spuriously projects sensible qualities into a
domain where such qualities have no place. In a manner of speaking, reification 'corporealizes' what is inherently incorporeal and
thus confounds the physical with the corporeal plane.

It cannot be denied that reification was rife throughout the Newtonian era. There was first of all mechanics of rigid and nonrigid
bodies, of subcorporeal objects, therefore, which were no doubt
routinely reified through identification with the corresponding corporeal entities. There was also gravitation, to be sure, which could
not be dealt with in this way; but this circumstance was perceived as
an anomaly. Newton himself tried (in the *Opticks*) to explain gravitational force in terms of the pressure gradient of a hypothetical
interplanetary fluid; but he also recognized with admirable clarity
that in a technical or computational sense the question had no
bearing on physics at all. To calculate the motion of bodies under
the action of gravitational force, what alone matters is the mathematical law that describes how one 'particle of mass' affects another;
and Newton had good reason to suppose that his own law of gravitation had settled this matter once and for all.

The hankering for mechanistic explanations, however, did not
abate. It was an age when men of science looked expectantly upon
Mechanics as the key to the unraveling of just about every phenomenon; and that *Weltanschauung*, as we know, did have its victories.
In addition to his prime discoveries—the laws of motion and of
gravity, and the consequent explanation of planetary orbits—
Newton himself pioneered an acoustics which in effect reduced
sound to a phenomenon of continuum mechanics, and began at
least to speculate—quite correctly—that temperature and heat had
to do with the 'vibratory agitation of particles'. It is not without
interest to note that a second theory of heat, less felicitous but no
less mechanical than the Newtonian, made its appearance at

approximately the same time and was widely accepted for about two hundred years. According to this view heat was supposedly a 'subtle, invisible and weightless' fluid named phlogiston, which was thought somehow to permeate bodies and flow from hot regions to cold, much as ordinary fluids flow along a pressure gradient. Not until the middle of the nineteenth century was the phlogiston doctrine finally abandoned in favor of the Newtonian theory, thanks to the work of Joule and Helmholtz.

Apart from the various branches of mechanics—including the still problematic theory of heat—Newtonian physics also comprised optics as a more or less independent and successful branch of inquiry. No one had any serious doubts that this domain, too, could eventually be understood in mechanical terms, and there existed in fact two mechanistic models—the wave model of Huygens and the corpuscular theory of Newton—purporting to explain the phenomenon of light.

There was also a rudimentary chemistry, to which Newton, for one, devoted an immense effort. Only it happens that there was not the slightest possibility at the time of explaining chemical phenomena in mathematical, let alone mechanical terms—which is no doubt the reason why Newton never published a separate treatise on this subject. As was however to be expected, Newton and his peers inclined strongly towards a mechanistic theory of atoms, which soon came to be regarded in wider circles as an incontrovertible dogma of science. As Voltaire has put it, with his customary aplomb:

> Bodies most hard are looked upon as full of holes like sieves, and in fact this is what they are. Atoms are recognized, indivisible and unchangeable, principles to which is due the permanence of the different elements and of the different kinds of beings.[9]

It is to be noted, finally, that in addition to mechanics and optics—and to an imagined atomism—the Newtonians were also conversant with electric and magnetic phenomena of a rudimentary

9. See W.C. Dampier, *A History of Science* (Cambridge: Cambridge University Press, 1948), p167.

kind.[10] For various reasons, however, little progress could be made in this domain until the nineteenth century, when the requisite means became available and research prospered, culminating in the magnificent theory of Faraday and Maxwell. And with the unveiling of the electromagnetic field the mechanistic outlook began at last to wane. The concept of pure structure, of mathematical form, was about to supplant the mechanical notions of the Newtonian epoch. But the transition was gradual. Maxwell himself conceived of the electromagnetic field along mechanical lines in terms of an ether—another 'subtle, invisible and weightless' fluid not unlike the ill-fated phlogiston—and this view was widely accepted for several decades. In retrospect one can see that a powerful bias in favor of mechanistic explanations was still in effect within the scientific community, and that it apparently required the full force of refined experiment plus the bold genius of Einstein to overcome this inveterate propensity. The transition was, however, accomplished, and we have by now become reconciled to the electromagnetic field, for example, as a physical entity in its own right, a 'structure' which can not be reduced to mechanical categories.

But although we have jettisoned the ether and no longer hanker for mechanistic models, we still have need for sensible supports. The electromagnetic field, no less than any other physical object, is thus to be conceived—not, to be sure, in mechanistic terms—but yet by virtue of appropriate representations of a visual kind. As every student knows, the electric field at a point is given by a vector, a mathematical entity which has a length and direction and can consequently be pictured as an arrow—a small one, preferably, which can conveniently be localized at the point in question. One inclines, in fact, to position the arrow with its 'tail' exactly at the point P. With a little effort one can now picture an electric field at a given time as a continuous three-dimensional distribution of such

10. Not only did Newton recognize gravitational and electromagnetic force, but it appears that he also anticipated nuclear forces, as one may gather from the following statement in the 31st Query of the *Opticks*: 'The attractions of gravity, magnetism and electricity, reach to very sensible distances, and so have been observed by vulgar eyes, and there may be others which reach to so small distances as hitherto to escape observation.'

arrows, which change their length and direction in accordance with the demands of the mathematical theory. The same can be done for the magnetic field, and hence for the electromagnetic, which thus requires the attachment of *two* arrows to each point, corresponding to the electric and magnetic components of the field. To further facilitate our comprehension one could even think of the electric vectors as red and the magnetic as blue, a device which enables one to produce impressive pictures of an electromagnetic wave.[11] I am not suggesting, of course, that anyone could be simple enough to take at face value the notion of 'red and blue vectors'; my point, rather, is twofold. Firstly, it is to be admitted that at least on a mental plane representations of this general kind are necessary and indeed legitimate as a sensible support for the concept of an electromagnetic field. And this being the case, it is in principle possible— and indeed quite easy—to reify the electromagnetic field; all that one needs to do in that regard is to forget that an electric or magnetic vector at P is not in fact an arrow, but something of a totally different kind which indeed can not be 'pictured' at all—except of course by way of an artifice, such as that of the arrow. In a word, there is a leap to be made—and it may not be easy to tell from the outside whether a person 'is looking at the finger or at the moon.'

One could argue that from a sufficiently pragmatic point of view it hardly matters; and this is often true. In this instance, however, it happens that the indicated reification of the electromagnetic field is inadmissible even from a technical standpoint, due to the fact that the electric and magnetic vectors are not Lorentz invariant. The decomposition of the electromagnetic field into electric and magnetic components, in other words, depends upon the choice of reference frame. And what alone is invariant and hence objectively real turns out to be, not a pair of vectors in three-dimensional space, but a so-called exterior 2-form in a four-dimensional space-time. Meanwhile our 'red and blue vectors' retain nonetheless their validity and use as a representation of the electromagnetic field—so long as it is understood that such a picture is not to be taken at face

11. One needs of course to take into account the time dependence of the field. This can be done, for example, through animated graphic display.

value, and that even in a formal sense it applies only within a restricted class of reference frames. So far as the exterior 2-form is concerned, this too stands in need of visual supports; but there exists no 'picture'—no single concrete representation in ordinary space and time—with which this mathematical object could be identified. In a word, the electromagnetic field cannot be reified in a Lorentz-invariant way.

The same applies in fact to other Lorentz-invariant structures, and thus to relativistic physics as a whole. And this is no doubt the prime reason why relativity strikes us as quite so formidable: it is 'difficult' by virtue of the fact that it cannot be reified with impunity. When it comes to the microworld, moreover, the same happens even when the requirement of Lorentz invariance is neglected, inasmuch as the wave-particle dualism evidently prohibits reification of so-called particles. For indeed, these objects cannot be consistently pictured as particles, because in the context of certain experiments they behave as waves; and by the same token, they cannot be pictured as waves. Consequently they cannot be pictured at all—and this is precisely what puzzles us.

What has happened in our century is that physics has been driven on its own ground to reject naïve interpretations and maintain a rigorously symbolist stance in respect to concrete representations. Or better said, it has been forced to maintain such a stance in the domain of high velocities, and above all, in the microworld. When it comes to the ordinary macroscopic physical domain, on the other hand, the tendency to reify still manifests itself, even in authors who expostulate at length on the subject of 'quantum strangeness'—as if 10^{24} atoms could be pictured more readily than one! It has yet to be recognized that there is an ontological difference between the physical and the corporeal domains, and that the gap cannot be closed through the mere aggregation of so-called particles.

III

MICROWORLD
AND INDETERMINACY

It is one thing to speak of a generic physical object—such as 'the electromagnetic field,' for example—and quite another to speak of a specific physical object, the kind that exists concretely and can actually be observed. And the difference is this: Whereas the generic object is determined by a mathematical model or representation alone, the latter is subject, in addition, to determinations of an empirical kind. It is an object, in other words, with which we have already established a certain observational contact. For example, one can speak of the planet Jupiter because it has in fact been viewed or detected; and again, one could search for the planet Pluto (discovered in 1930) because the latter also had already been observed, not directly, to be sure, but by way of its effects upon other planets.

There are, of course, degrees of specification; the distinction, however, between the generic and the specific is nonetheless well defined, and turns out to be crucial. For it happens that physics deals, first and foremost, with physical objects of the 'specific' kind: these are its 'actual' objects, one could say, as distinguished from entities (such as 'the electromagnetic field') which exist in some abstract, idealized or purely mathematical sense. The 'actual' objects of physics, therefore, are entities which not only can be observed in some suitable sense, but have in fact been already observed. Like Jupiter or Pluto, they have been specified to some extent by a set of observations. I shall use the term 'specification' to refer to the empirical act or acts by which a physical object is specified; and with this understanding it can indeed be said that an object is not specific until it has been specified.[1]

Let us now consider a few examples of specification. In the case of subcorporeal objects it is normal or natural to specify SX by way of the corresponding corporeal object X, which is to say, by means of presentation. On the other hand, it is also possible to specify a subcorporeal object SX by more indirect means—as in the previously cited case of Pluto, for instance. Having been specified by whatever means, the object can of course be further specified through additional determinations; specification, as already said, is subject to degrees.

While subcorporeal objects can indeed be specified by presentation (or better said, by presentation alone), this option does not exist in the case of a transcorporeal object, such as an atom, for instance, or an elementary particle. Thus, when it comes to transcorporeal objects, specification necessarily proceeds in two stages: first, the object must interact with a subcorporeal entity, which in turn is observed (or rendered observable) through presentation. Consider, for instance, an electromagnetic field produced in the laboratory: in the first place the field interacts with the scientific apparatus by which it is generated; and this apparatus (conceived now as a subcorporeal object) can then be observed by way of presentation. Or again, a Geiger counter registers the presence (within its chamber) of a charged particle. The particle enters the chamber and causes an electrical discharge, which then is registered somehow on the corporeal level (in the form of a audible click, perhaps, or the reading of a counter). Now this chain of events constitutes, evidently, a specification of the particle. One may henceforth speak of 'particle X'—even though it may not be possible ever again to re-establish observational contact with particle X. On the other

---

1. This does not necessarily mean, however, that a specific physical object did not exist prior to its specification. I am not suggesting, for example, that the planet Jupiter materialized somehow at the moment when it was first observed. What I am saying is that one needs first of all to specify an object before one can ask, among other things, whether that object existed, let us say, a thousand years ago. And in the case of Jupiter, of course, the answer to this question turns out to be affirmative. There are other kinds of objects, as we shall presently see, where this happens not to be the case.

hand, with the aid of more complicated instrumentation the experimentalist is able not only to establish an initial observational contact with a particle, but can follow up with additional observations. Having specified 'particle X,' in other words, he can subject this particle to further measurements—as was done, for example, by Hans Dehmelt, the recent Nobel laureate, who managed to 'imprison' a positron in a so-called Penning trap over a period of some three months, during which the given particle (dubbed 'Priscilla') could be observed to unprecedented degrees of accuracy.

But be that as it may, what presently concerns us is the following general fact: Whether one deals with a fundamental particle or with the simplest subcorporeal entity, one cannot speak of a physical object X until a certain initial observational contact with X has been established. Physical objects do not simply 'grow on trees': they need first of all to be 'specified' in the technical sense which we have attached to this term.

❖   ❖   ❖

The question arises now whether it is possible to specify a physical object so thoroughly that the outcome of all additional observations can be predicted, or is in any case determined in advance. It will be expedient, however, to rephrase this question slightly, after introducing some further distinctions. In conformity with accepted usage, I shall use the term 'system' to designate an abstract or mathematical representation of a physical object. A physical object, conceived in terms of a given representation, can then be termed a physical system. It is the representation or abstract system, moreover, that defines the observables: the quantities associated with the physical system, which can in principle be determined by empirical means. What is and what is not an observable, in other words, depends, not simply upon the object, but upon the way in which that object is conceived. A billiard ball, for example, regarded as a rigid sphere, admits an indefinite number of rather unsophisticated observables (beginning with its mass, its diameter, and its position and velocity coordinates); conceived as an ensemble of atoms, on the other hand, it admits a host of other observables. Specification refers consequently to the physical system, as distinguished from the object as such. Given a physical system and a subset of its observables, one can say that this subset is specifiable if it is possible to measure each observable in the subset (so that, at the termination of the composite experiment, the values of all these observables are known). The question, posed above, can therefore be restated as follows: Given a physical system, does there exist a specifiable subset of its observables, the experimental determination of which will determine the values of all other observables of the system? Is it possible, in other words, to render a physical system fully determinate by way of specification? One knows today, in light of quantum theory, that this question is to be answered in the negative. There is in reality no such thing as a fully determinate physical system (one for which the exact values of all observables can be predicted). And this is so, not simply because one is unable to control or monitor external forces with the requisite precision, but equally on account of a certain residual indeterminacy intrinsic to the physical system itself, which no amount of specification can dispel.

On the other hand, so long as one is dealing with large-scale physical systems of a sufficiently simple kind, the effects of this residual indeterminacy may not be measurable, or may be so small as to play no significant role.[2] In a formal and approximative sense, therefore, one can speak of a determinate physical system; and these, to be sure, are precisely the systems with which classical physics is concerned, and to which it applies. Such a system can then be described or represented in terms of a complete set of observables— a set in terms of which all other observables can be expressed. And this means that we need no longer distinguish between the system as such and its observables; the system can be identified, in effect, with a complete set of observables. What, for example, is an electric field, classically conceived? It is a continuous distribution of electric vectors: of observables, that is! Such a reduction of the system to a

---

2. Strictly speaking, it is not just the number of atoms, let us say, that counts in that regard, but also the arrangement of these atoms. In the case of so-called aperiodic arrangements, for example, quantum effects may come into play even for macroscopic ensembles.

subset of its observables, moreover, is in fact implied by the very formalism of pre-quantum physics, which deals exclusively with functional relationships between observable quantities. A classical physical system, thus, is nothing more than a distribution in space and time of certain observable scalar or tensor magnitudes.[3]

Where there is indeterminacy, on the other hand, the classical formalism breaks down. One needs then to distinguish categorically between the physical system S and its observables, not all of which can in principle be determined through specification. The classical reduction (of the system to its observables) is consequently admissible only in what may be termed the classical limit: under conditions, that is, which guarantee that the effects of indeterminacy play no measurable or significant role. Outside this limit, or this restricted domain, physics requires a non-classical formalism—a need which was brilliantly met in 1925 with the discovery of quantum mechanics. The new formalism, as we know, distinguishes between system and observables, and on this basis enables one to transact the business of physics in the face of indeterminacy.

❖   ❖   ❖

A distinction is often made between the so-called microworld and macroworld—as if the physical universe could somehow be split up into two subdomains answering respectively to these designations. One might ask, of course, just how many atoms or subatomic particles are required to take us from the microworld into the macroworld; but then, what is the point of this distinction in the first place? Now the point, it seems, is that 'large scale' or so-called macrosystems are supposed to lend themselves to description in more or less 'continuous' terms. They consist thus of atomic or subatomic aggregates which can be effectively approximated by classical models. It needs however to be clearly understood that the distinction

3. It is reasonable to surmise that this 'passage to the classical limit' may not be legitimate in the case of even the simplest living organisms. As some have conjectured, it is not unlikely that quantum indeterminacy plays a vital role in the phenomena of the biosphere.

between 'large' and 'small' aggregates is void of any ontological significance. Or to put it another way: the notion of macrosystem, in particular, belongs to the practical or pragmatic realm; it has to do with degrees of approximation and the feasibility of certain simplified models. In reality, however, every physical object constitutes a microsystem—by virtue of the fact that it is composed of atoms or fundamental particles. The microworld, thus, so far from constituting a subdomain, coincides actually with the physical universe in its totality.

Scale, meanwhile, does have its significance. The point, however, is not that physical reality becomes somehow strange 'in the small,' but that one is obliged, as one moves in the direction of the small, to discard idealized models and deal eventually with the physical object as an aggregate of fundamental particles. The circumstance that physical objects without exception are in truth composed of these smallest so-called particles signifies that the physics of these 'particles' is indeed the fundamental physics. It is thus in the atomic and subatomic domains, precisely, that physics is forced to descend, as it were, to its own fundamental level.

Nonetheless the belief persists that the physical universe does become increasingly 'strange' as one approaches atomic and subatomic dimensions. Large objects, supposedly, behave in a more or less familiar and reasonable manner, whereas atoms and particles act in ways that are most bizarre. So bizarre, in fact, that according to some authorities even the accustomed laws of logic cease to apply within that uncanny domain. It follows, however, from what has been said above, that the so-called large-scale objects of physics are in reality just as 'strange' as the electron or quark; it is only that when it comes to the former it is frequently permissible to ignore this strangeness, so to speak, by conceiving the object in terms of a classical model—the kind which does indeed answer more or less to the demands of our imagination or common sense. What is thus in a way familiar, however, is precisely the model and not the object as such. And one might add that even the former commends itself to our imagination only because one forthwith takes a second step: in one way or another one identifies the classical model with a corporeal object of some kind; in a word, after passing to the classical limit, one reifies. And then, at last—having safely returned, as it were, to the *terra firma* of the corporeal domain—one encounters the familiar; for indeed, to us the familiar is none other than the perceptible.

Meanwhile the microworld—and thus the physical universe at large—is admittedly 'strange' in the sense that it can be neither perceived nor imagined; but it is not 'quantum strange' in the more or less popular sense. For example, it is by no means the case that the electron is sometimes a particle and sometimes a wave, or that it is somehow particle and wave at once, or that it 'jumps' erratically from point to point, and so on. For indeed, this kind of 'quantum strangeness' stems quite simply from a failure to distinguish between the microsystem as such and its observables (the electron, in this instance, and its position, momentum, and other dynamic variables). In effect, one treats the latter as classical attributes of the electron, which they are not, and cannot be. Or to put it another way, one spuriously projects the results of distinct and interfering measurements upon the electron itself, which consequently seems to combine logically incompatible attributes. It is thus that the electron may appear to be both wave and particle, or to engage in a regimen of 'jumping' which does indeed defy comprehension. One could say that this kind of 'quantum strangeness' results from an uncritical and spurious realism—a realism which in effect confounds the physical and the corporeal planes.

The prevailing Copenhagen interpretation, on the other hand, avoids this pitfall by eschewing realism altogether in regard to the microworld. 'There is no quantum world,' said Bohr; and whereas there has been considerable debate as to what precisely Bohr meant by this oft-quoted dictum, Copenhagenists as a rule shy away from an overtly realist conception of microphysical systems. Their dominant tendency, it would seem, is to keep out of trouble, so to speak, by resorting to a basically positivistic stance when it comes to the microworld.

For us, on the other hand, the microworld is objectively real—as real, indeed, as the physical universe at large, with which in fact it coincides.

❖   ❖   ❖

It has often been said that the microworld is indeterministic,[4] and this claim is based, presumably, on the Heisenberg uncertainty principle, or on the phenomenon of indeterminacy, which amounts to the same. The question remains, however, whether Heisenberg uncertainty—or indeterminacy—implies indeterminism.

To begin with, let us note that Heisenberg uncertainty refers, not to the microworld or physical universe as such, but to the results of measurements, and thus to a transition from the physical to the corporeal plane. On the plane of the microworld itself, on the other hand, there is no such thing as Heisenberg uncertainty. One cannot say, for example, that the position or the momentum of an electron is uncertain or indeterminate, for the simple reason that an electron—in and by itself—has no position, and no momentum as well. In technical parlance, it is described by a state vector, which as a rule will not be an eigenvector of either observable.

What, then, does the so-called state vector of a physical system tell us in general about an observable? It tells us primarily two things, both of which are probabilistic and consequently statistical in their empirical content. Thus, in the first place, the state vector determines an expected value, which is to say, the average value of the observable over a sufficiently large number of observations—a concept which can indeed be interpreted in precise terms. And in second place, the state vector determines a so-called standard deviation, another probabilistic quantity, which tells us, roughly speaking, how close, on the average, the observed values will be to the expected. And this notion, needless to say, can once again be given a precise statistical sense.

Now, let us recall that the Heisenberg uncertainty principle has to do with the standard deviations Δp and Δq associated with conjugate observables p and q. What the principle affirms, in fact, is that

$$\Delta p \ \Delta q \geq h/2\pi,$$

4. There is of course the classical determinism to be accounted for; but the problem is readily solved on the grounds that the classical laws which enable one to predict the evolution of a physical system are inherently probabilistic, and applicable only to the macroworld.

where h is Planck's constant. This constitutes a precise mathematical statement, which can be derived from the axioms of quantum theory, and interpreted empirically in terms of statistical ensembles.

What quantum theory hinges upon is the fact that the state vector—or equivalently, the physical system—though it does not, in general, determine the results of individual measurements, does in any event determine their statistical distribution. Meanwhile, however, there is absolutely nothing 'uncertain' about the physical system as such. The case is indeed analogous to that of a coin, which can come up 'heads' or 'tails' when tossed. Here, too, the fact that one cannot tell beforehand which way the coin will come up does not mean that the coin itself is somehow 'indeterminate'; the so-called uncertainty, in other words, pertains obviously to the toss, and not to the coin. And let us add that the latter—no less than a quantum mechanical system—determines the probability distribution of its 'observables.' For example, it determines the distribution (and thus the expected value and standard deviation) of the number of 'heads' in n trials—as every student of probability theory will recall.

If quantum mechanical systems, then, are not in themselves 'uncertain,' are they nonetheless indeterministic? Now, to say that a physical system is deterministic is to affirm, presumably, that the evolution of the system is uniquely determined by its initial state (assuming, of course, that we know the external forces impinging upon the system). But this is precisely what the celebrated Schrödinger equation implies! The microworld, thus, is indeed deterministic, even though physical systems are indeterminate. One might put it this way: The initial state of an isolated physical system (or of a physical system subject to known external forces) does determine its future states; but it happens that the state of a system does not in general determine the values of its observables. There is thus no conflict between determinism and indeterminacy; and as a matter of fact, quantum theory insists upon both. To be precise, it is the Schrödinger equation that guarantees determinism, even as the Heisenberg principle guarantees indeterminacy.

The objection may be raised that measurement destroys determinism; for as one knows, a measurement performed on a physical system can cause the so-called collapse of the state vector, an event which violates the Schrödinger equation. One could say that measurement abolishes determinism by interrupting the 'normal' evolution of the physical system. It is to be recalled, however, that physical systems are specified by way of measurement. Insofar, therefore, as a measurement collapses the state vector, it constitutes an act of specification which alters the state, and thus the 'actual' physical system. The physical system X with which we were concerned prior to the measurement will not in general be the same as the system Y resulting from this additional specification. So long as one is dealing with determinate physical systems, of course, the system can be specified once and for all. There is then no collapse of the state vector and no change of specification—or 'loss of identity'—resulting from subsequent acts of measurement. When it comes to indeterminate systems, on the other hand, subsequent measurements will in general result in the specification of a new physical system. One could say that the original physical system is terminated—or metamorphosed—by the collapse of its state vector. To be sure, quantum mechanical systems are not perdurable, nor are they 'absolute'—but exist 'for us,' as objects of intentionality. These basic facts, however, do not impede determinism, the point being that a quantum mechanical system behaves nonetheless in a deterministic way (so long as it exists).

Obviously enough, this quantum mechanical determinism is a far cry from the classical. However, what has been forfeited is not so much determinism as it is reductionism: the classical supposition, namely, that the corporeal world is 'nothing but' the physical. It is this axiom that has in effect become outmoded through the quantum mechanical separation of the physical system and its observables. Quantum physics, as we have seen, operates perforce on two planes: the physical and the empirical; or better said, the physical and the corporeal, for it must be recalled that measurement and display terminate necessarily on the corporeal plane. There are, then, these two ontological planes, and there is a transition from the physical to the corporeal resulting in the collapse of the state vector. The collapse, one could say, betokens—not an indeterminism on the physical level—but a discontinuity, precisely, between the physical and the corporeal planes.

But whereas the very formalism of quantum mechanics proclaims that there are these two levels and cries out, as it were, for the recognition of this fact, the prevailing reductionist bias has impeded that recognition from taking place. It is little wonder, therefore, that the ontological interpretation of quantum mechanics has not gone well.

❖   ❖   ❖

Quantum mechanics suggests that microphysical systems constitute a kind of potency in relation to the actual world. As Heisenberg points out, they occupy in effect an intermediary position between non-existence and actuality, and in this respect are reminiscent of the so-called Aristotelian *potentiae*.

To understand this more clearly, we need to take a somewhat closer look at the quantum mechanical formalism. Let us note, first of all, that every observable admits a set of possible values (its so-called eigenvalues), and that in general a measurement of a given observable can yield any of these admissible results. A physical system, however, can also be in a state in which the value of the given observable is determined with certainty; and these states are called eigenstates. For example, if a measurement of the observable yields the eigenvalue λ, then the system, at that moment, is known to be in an eigenstate corresponding to λ.[5]

I have already alluded to the fact that a physical system, quantum mechanically conceived, is represented by a so-called state vector. More precisely, state vectors represent *states* of a physical system.[6] And this evidently explains the notion of eigenvectors to which I also referred (in the discussion of indeterminacy): an eigenvector, thus, is a state vector corresponding to an eigenstate.

5. We are assuming that the measurement is effected by an experiment 'of the first kind'. There are also experiments 'of the second kind' which do not leave the system in a corresponding eigenstate.

6. It should be said that a state vector can be multiplied by a complex number, and that multiplication by a nonzero factor does not alter the corresponding physical state.

Now, it will be recalled that vectors can be added, and also multiplied by a number (real or complex, as the case may be); and this means that vectors can be combined to form weighted sums. Thus, every weighted sum of state vectors (so long as it is not zero) defines another state vector.[7] Since state vectors, however, represent states of the physical system, every such weighted sum corresponds to a physical state. One arrives thus at the so-called superposition principle, which affirms that weighted sums of state vectors correspond to an actual superposition of states. It turns out, in other words, that the algebraic operations by which one forms weighted sums of state vectors (with complex coefficients, no less) carry a physical significance. There exists, if you will, an 'algebra of states,' which permits us to represent physical states in multiple ways as a superposition of other states.[8]

7. The weights or coefficients in these weighted sums are in general complex numbers, and this fact is vital to quantum theory. If we did not have complex numbers at our disposal (numbers which involve the 'imaginary' square root of −1), we would be unable to understand the microworld.

8. The superposition of quantum mechanical states can be understood by analogy to the superposition of sound waves. Consider a tone produced by a musical instrument: a violin, an oboe, an organ, and so forth. Each of these tones has its own characteristic, its own timbre, as it is called; and that is why we can recognize the instrument from its tone. Each tone, however, can be represented as a superposition of so-called pure tones: tones, namely, whose sound wave is a simple sinusoid. And that is what an electronic synthesizer does: it produces the sound of a flute, for instance, by mixing a number of pure tones in the right proportions. Another example of superposition is provided by the fact that an arbitrary color can be obtained as a superposition of three primary colors. Or again: white light, when passed through a prism, breaks up into light of various colors (a process which can again be reversed). It is to be noted, moreover, that in all these instances of superposition we are dealing ostensibly with wave motion of one kind or another. Now, inasmuch as superposition is fundamental to quantum mechanics, and appears to be a wave phenomenon, one is led to surmise that quantum entities may in fact be waves; and this idea has indeed been seriously entertained by many physicists, beginning with Erwin Schrödinger (one of the founders of quantum theory). The reader may recall that the term 'wave mechanics' has often been used as a synonym of quantum theory. It must however be understood that if quantum entities are in fact 'waves', they are necessarily 'sub-empirical' waves: waves which in principle cannot be observed. For as we know, quantum theory insists that the physical system is one thing, and its observables another. It is not clear, therefore,

The question arises whether, for an arbitrary observable, every state of the system can be represented as a superposition of eigenstates. Can every state vector, in other words, be expressed as a weighted sum of eigenvectors belonging to the given observable? And whereas this happens not to be the case, one is able, in general, to obtain an analogous representation by mathematically more sophisticated means.[9] However, to avoid technical complications which have no bearing upon the argument, I will suppose that every observable does have a 'complete' set of eigenvectors: a set, namely, in terms of which every state vector can be expressed as a weighted sum.

Now, what does all this have to do with Heisenberg's contention to the effect that quantum systems constitute a kind of Aristotelian *potentia*? This is what must now be explained. Consider the representation of a state vector as a weighted sum of eigenvectors belonging to a given observable. Each eigenvector corresponds to an eigenstate, and thus to a possible outcome of an actual experiment. It thus represents a certain empirically realizable possibility, the probability of which is in fact determined by the weight with which that eigenvector occurs in the given sum.[10] The state vector itself, as a weighted sum of eigenvectors, may consequently be viewed as an ensemble or synthesis of the possibilities in question. And if one

---

that anything is really to be gained by speaking of quantum systems as 'waves'. In the final count, it appears that the superposition principle tells us all that can and all that need be said on the matter. It affirms, if you will, that quantum entities can be superposed 'as if they were waves of some kind.' And let us add, for readers with a certain exposure to the mathematics of quantum theory, that the ubiquitous phase factor $\exp(-2\pi i Et/h)$ on the level of state vectors does indeed testify to the 'wave nature' of quantum states. Quantum theory, one can say, has in effect resolved the wave-particle quandary by relegating the two mutually contradictory concepts to different ontological planes: waves to the physical, and particles to the empirical, that is to say, the corporeal plane. That, in any case, is what the quantum-mechanical separation of the system and its observables effects *de jure*, even if people, *de facto*, continue to confuse the issue by confounding the physical with the corporeal domain.

9. In place of eigenvectors one must use what Dirac terms 'eigenbras'; and in place of finite or infinite sums, one requires integrals of an appropriate kind.

10. Assuming that the sum of the squared absolute values of the weights equals 1 (a condition that can always be achieved by multiplying the state vector by a

assumes (as we have done) that the state vector can be expressed as a weighted sum of eigenvectors for each and every observable, it then constitutes, by the same token, a synthesis of *all* possible outcomes for every conceivable measurement that can be performed on the given physical system.[11]

At the termination of a measurement, on the other hand, the system will be in an eigenstate belonging to the given observable. If the state vector, prior to measurement, was a weighted sum of eigenvectors, it is now a particular eigenvector, and thus, if you will, a weighted sum of eigenvectors in which all coefficients but one are zero. The state vector has collapsed, as we say; in an instant it has been reduced to a single eigenvector of the given observable: a single possibility, that is, the probability of which has now jumped to the value 1 (indicative of certainty). By the act of measurement a particular element from the given ensemble of possibilities has been singled out and realized on the empirical, that is to say, the corporeal level. The physical system, as an ensemble of possibilities, has thus been 'actualized.' But only in part! For whereas the value of a particular observable has now been determined, the system remains in a superposition of eigenstates for most other observables. And therefore, despite partial actualizations effected by measurement, the system is and remains an ensemble or synthesis of possibilities. In the words of Heisenberg, it is not in reality a 'thing or fact,' but rather a potency, a kind of *potentia*.

As the Aristotelian terminology itself suggests, the conception of physical systems and state vector collapse at which we have arrived is in a way classical, and can in fact be understood from a traditional metaphysical point of view. It has long been known that the transition from the possible to the actual—or from potency to

---

suitable nonzero factor) and that there are no multiple eigenvalues, the probability that a measurement will realize the possibility corresponding to a particular eigenvector is given by the squared absolute value of the corresponding weight.

11. When I speak of a state vector as 'an ensemble of possibilities', I am in effect identifying the state vector with the corresponding physical state. Strictly speaking, it is of course the physical system in a given state (and not its mathematical representation!) that is 'an ensemble or synthesis of empirically realizable possibilities'.

manifestation—entails invariably an act of determination: a choice of one particular outcome out of an ensemble of possibilities. Euclidean geometry, moreover, exemplifies this process very clearly—but only so long as the discipline is understood in the ancient way. One must remember that prior to Descartes the geometric continuum—the Euclidean plane, for example—was conceived as an entity in its own right, and not simply as the totality of its points. According to the pre-Cartesian view, there *are* in fact no points in the plane—until, that is, they are brought into existence through geometric construction. Classically conceived, the plane as such is void; in itself it constitutes a kind of emptiness, a mere potency, in which nothing has yet been actualized. And then one constructs a point or a line, followed by other geometric elements, until a certain figure is obtained. It is to be noted that these determinations cannot actually be made on rational grounds, or on the basis of some prescribed rule, a circumstance which tends to puzzle the analytic mind. The determinative act, moreover, is in fact more than a mere choice, a mere selection of one element from a given ensemble: for it brings into existence—as it were, *ex nihilo*—something which previously did not exist as an actual entity. Geometric construction, classically conceived, is thus suggestive of cosmogenesis. One could say that it imitates or exemplifies the creative act itself within the mathematical domain.

Getting back to quantum mechanics, and in particular, to the act of measurement, one now perceives that this can indeed be interpreted in traditional ontological terms. Measurement, thus, is the actualization of a certain potency. Now the potency in question is represented by the (uncollapsed) state vector, which contains within itself, as we have seen, the full spectrum of possibilities to be realized through measurement. To measure is thus to determine; and this determination, moreover, is realized on the corporeal plane: in the state of a corporeal instrument, to be exact. Below the corporeal level we are dealing with possibilities or *potentiae*, whereas the actualization of these *potentiae* is achieved on the corporeal plane. We do not know how this transition comes about.[12]

---

12. We shall return to this question in chaps. 5 and 6.

Somehow a determination—a choice of one particular outcome from a spectrum of possibilities—is effected. We know not whether this happens by chance or by design; what we know is that somehow the die is cast. And this 'casting of the die' constitutes indeed the decisive act: it is thus that the physical system fulfills its role as a potency in relation to the corporeal domain.

❖   ❖   ❖

A word, now, concerning the superposition principle. Dirac was perhaps the first to observe that the principle has no analogue in the classical domain. It is true that solutions of a linear homogeneous equation can be 'superimposed,' and this fact underlies the Fourier analysis, for example, of classical vibratory systems. But as Dirac has made clear, 'The superposition that occurs in quantum mechanics is of an essentially different nature from any occurring in the classical theory, as is shown by the fact that the quantum superposition principle demands indeterminacy in the results of observations in order to be capable of a sensible physical interpretation.'[13] The superposition principle, thus, applies necessarily to a level of reality on which the values of observables have not yet been fixed: to the microworld, namely, which is a realm of potency, a sub-actual domain. The transition to actuality must therefore involve a certain 'de-superposition'—which is none other than the collapse of the state vector.

There is nothing in the state vector itself that could explain or account for this determinative act—even as there is nothing in the Euclidean plane that would permit us, by some kind of rule, to pick out a point or a line. On the other hand, inasmuch as the act of measurement entails an interaction with a second system, it is hardly surprising that the first, by itself, does not suffice to explain state vector collapse. What has, however, puzzled physicists, is that even with the second system in place, one is no better off: for it happens that the combined system is again in a superposition of

13. *The Principles of Quantum Mechanics* (Oxford: Oxford University Press, 1958), p14.

eigenstates for the given observable. But disturbing or paradoxical as this fact may be so long as one fails to distinguish between the physical and the corporeal planes, it is exactly what one would expect once this fundamental distinction has been recognized. The point is that the transition from potency to actuality requires invariably a creative act—a creative fiat, one could say—which nothing in the domain of potency can account for or explain. Nothing within the physical plane, therefore, could cause a state vector to collapse—distressing as this fact may be to those who imagine that there is nothing beyond the physical.

These considerations, admittedly, do not resolve the so-called measurement problem; they do, however, make clear why current attempts to find a solution have failed. I will defer to a later chapter the question whether quantum mechanics constitutes a 'complete' theory or not; the point that concerns us presently is that quantum mechanics could at best be a complete theory of the *physical* universe. For it stands to reason that so long as the corporeal order does not reduce to the physical, neither quantum mechanics nor any other physical theory could be 'complete' in an unrestricted sense. It is only to be expected, therefore, that a well-formulated physical theory will somehow bear witness to this ontological limitation. And thus one need not be surprised that the Schrödinger evolution of physical systems should display 'gaps' which quantum mechanics itself cannot predict, and that these appear precisely when it comes to the fateful transition which takes us out of the physical plane. So far from being indicative of imprecision, this basic feature of quantum mechanics testifies rather to its correctness and sufficiency. The seeming completeness of classical physics, on the other hand, betokens the fact that we are dealing, not so much with physical realities, as with convenient abstractions. There is point, after all, to Whitehead's provocative precept: 'Exactness is a fake.'

Getting back to the superposition principle, it is to be noted that in the case of a subcorporeal system certain superpositions are evidently ruled out. In the case of a scientific instrument, for example, a pointer cannot be in two distinguishable positions at the same time. Thus, for any subcorporeal system SX, it must be assumed that only states which are 'perceptually indistinguishable' may conceivably be superposed. The point, clearly, is that the subcorporeal object is partially actualized by presentation; and actualization, as always, entails a determination, and thus a de-superposition.

It is of interest that this recognition at once resolves the so-called Schrödinger cat paradox. A single radioactive atom, if you will, is placed near a Geiger counter. If the atom disintegrates, it triggers the counter, which in turn sets up a certain chain of events ending in the untimely death of Schrödinger's cat. Now, inasmuch as the atom is admittedly in a superposition of states (disintegrated and undisintegrated), one reasons that the Geiger counter and the cat must likewise be in a corresponding superposition. And this would be rigorously true, moreover, if the counter and cat were quantum systems and nothing more. But it happens that both systems are subcorporeal, and that the superposition in question is of the kind that is ruled out: it is not possible for a Geiger counter to 'click' and not to 'click' within a given interval of time, nor is it possible for a cat to be both dead and alive at the end of the experiment. If the (normalized) state vector of the atom is of the form

$$0.6|\psi_1\rangle + 0.8|\psi_2\rangle,$$

for instance, where $|\psi_1\rangle$ and $|\psi_2\rangle$ correspond to the disintegrated and undisintegrated states, respectively, this does not imply that the state vector of the cat is in a corresponding superposition: it does not mean that the cat is 36% dead and 64% alive.[14] What it does mean is that it has a 64% chance of survival—a fact which needs, of course, to be interpreted in statistical terms.

There is no exceptional mystery here. Nor is it necessary (as some have suggested) to open the hatch and peek at the hapless cat in order to collapse its state vector. The cat collapses its own state vector, one might say, by the fact that it exists on the corporeal plane.

_____

14. According to quantum theory, the probability that a measurement will collapse a (normalized) state vector to a given eigenvector equals the squared absolute value of the corresponding coefficient. One arrives thus at the probabilities .36 and .64, corresponding to the eigenvectors $|\psi_1\rangle$ and) $|\psi_2\rangle$, respectively.

❖   ❖   ❖

As noted earlier, the frequent claim to the effect that the microworld is indeterministic—or somehow vague and fuzzy—reposes ultimately upon a confusion between the physical and the corporeal domains. The fact, for example, that the position and momentum of an electron cannot both be accurately ascertained is taken by the proponents of indeterminism to signify that the electron itself is ill-defined, or subject to random behavior. One quite forgets that the particle—the physical system, namely—is one thing, and its observables another. One forgets, in other words, that the electron as such *has no* position and no momentum—unless, of course, it happens to be in an eigenstate of the observable in question. Meanwhile, however, this so-called particle is neither vague nor fuzzy, nor indeed does it jump about in some bizarre and random fashion. Of all the things, in fact, with which physics has to deal, there is nothing more sharply defined and accurately knowable than the electron.

Mention should be made, in this connection, of its so-called static attributes, such as mass, charge and spin. Unlike the dynamic attributes—which, as we have seen, are not attributes at all—these quantities do belong to the electron as such. And they are measurable with stupendous accuracy. Recent measurements of the magnetic moment, for instance, have led to the value 1.001 159 652 188 (in appropriate units), with a possible error of 4 in the last digit.[15] As Richard Feynman has pointed out: 'If you were to measure the distance from Los Angeles to New York to this accuracy, it would be exact to the thickness of a human hair.'[16] Moreover, this magnetic moment can also be calculated by way of quantum electrodynamics; the answer appears then as the sum of a convergent infinite series, in which the successive terms decrease rapidly but become progressively more laborious to evaluate. And whereas calculations completed up to this time have not yet been able to match the accuracy

15. Hans Dehmelt, 'A single atomic particle forever floating at rest in free space,' *Physica Scripta*, T22 (1988), p102.
16. *QED: The Strange Theory of Light and Matter* (Princeton: Princeton University Press, 1988), p7.

of the latest experiments, they have indeed confirmed the digits 1.001 159 652. We know of no domain of physics in which agreement between theory and experiment has been more spectacular.

The fact is that physics comes into its own in the microworld, on the level of atoms and subatomic particles. It is here, precisely, that things become sharply defined. No longer, for instance, must one work with crude macroscopic parameters—such as the radius of a planet, or the density of this or that—but can deal instead with fundamental constants: the mass, charge, or magnetic moment of the electron, for example. The transition from classical to quantum mechanics, moreover, so far from complicating the formalism, amounts to a tremendous simplification; for indeed, the superposition principle brings into play what is actually the most manageable of all mathematical structures: to wit, a Hilbert space. Every mathematician understands full well what a luxury it is to work in a linear space; it is indeed to find oneself, mathematically speaking, in the best of all possible worlds. In short, one could say that the atomic and subatomic domains are 'made to order' for the physicist; it is here that one encounters the fundamental mathematical forms, unencumbered, so to speak, by accidental complexities.

But what are these fundamental 'forms'? None other, one is ultimately forced to reply, than the *bona fide* archetypes of the microworld, and thus of the physical universe at large. The principal goal or primary function of physics—according to this inherently Platonist view—is thus to ascend from the empirical domain to the level of mathematical archetypes. It is these that constitute its true objects, and not their fleeting reflections on the empirical plane.

But this ontological position is obviously at odds with the dominant spirit of our time. We are inclined to posit reality on the empirical plane and regard the mathematical forms—what Bohr refers to in the context of microphysics as the 'abstract quantum description'—as little more than artificial means for keeping track of empirical data.[17] For the nominalist, thus, it is the mathematical

---

17. And yet no one seems to be satisfied with this point of view. As I have noted before, physicists are not primarily concerned with positivistic games but would know the transcendent entities which reveal themselves in terms of measurable

form that somehow approximates the empirical data, whereas the Platonist, for his part, insists that the matter stands just the other way round: that it is the empirical data that reflects—and in a sense approximates—the mathematical form. It amounts to a question of ontological priority, of what comes first: the universal or the particular, the constant or the ephemeral.

However, it is in any case to be admitted that a realist stance vis-à-vis the microworld can be upheld on a Platonist basis alone. Atoms and subatomic particles can be 'real' only to the extent that mathematical forms are *bona fide* archetypes. As Heisenberg has put it: 'The "thing-in-itself" is for the atomic physicist, if he uses this concept at all, finally a mathematical structure.'[18]

Meanwhile it appears that the facts are definitely favorable to the Platonist contention. How else could one explain the stupendous success of mathematical physics?

❖   ❖   ❖

Among the many and various contemporary philosophies of physics, the closest by far to the position unfolded in this monograph is the philosophy of Werner Heisenberg. It may now be of interest to compare the two doctrines.

As is well known, Heisenberg considered himself a member of the Copenhagen school. At his hands, however, the so-called Copenhagen interpretation assumed a distinctive form, the salient feature of which lies in a realist view of the microworld, based upon

---

effects. In a word, they are 'realists' at heart. It is only that they often gravitate towards nominalistic premises that conflict with their realist intuitions. Here too, perhaps, one can speak of 'good physics' being unconsciously spoiled by 'bad philosophy.'

18. *Physics and Philosophy* (New York: Harper & Row, 1962), p91. Elsewhere, Heinsenberg has this to say: 'If we wish to compare the findings of contemporary particle physics with any earlier philosophy, it can only be with the philosophy of Plato; for the particles of present-day physics are representations of symmetry groups, so the quantum theory tells us, and to that extent they resemble the symmetrical bodies of the Platonic view.' See *Encounters with Einstein* (Princeton: Princeton University Press, 1989), p83.

the Aristotelian conception of potency. According to Heisenberg, there exist two ontological domains: 'In the experiments about atomic events we have to do with things and facts, with phenomena that are just as real as any phenomena of daily life. But the atoms or the elementary particles themselves are not as real; they form a world of potentialities or possibilities rather than one of things and facts.'[19] To deal with these two disparate domains, moreover, physics has need of two languages: the language of classical physics, in the first place, which applies to the world of 'facts and things'—and to the scientific instruments which are a part of this factual world—and the language of quantum mechanics, which applies to the domain of potentialities. In the state vector, interpreted *à la* Born as a kind of probability wave, Heisenberg perceives thus 'a quantitative version of the old concept of "potentia" in Aristotelian philosophy.'[20] It cannot be denied, of course, that a probability wave involves subjective elements; the salient feature of Heisenberg's philosophy, on the other hand, is his insistence that this probability wave entails also a 'completely objective' content—in the form of statements about *potentiae*, precisely.[21]

Quantum theory, thus, deals with two ontological domains; and the gap is spanned through measurement or observation:

> The transition from the 'possible' to the 'actual' takes place during the act of observation. If we want to describe what happens in the atomic event, we have to realize that the word 'happens' can apply only to observation, not to the state of affairs between two observations. It applies to the physical act of observation, and we may say that the transition from the 'possible' to the 'actual' takes place as soon as the interaction of the object with the measuring device, and thereby with the rest of the world, has come into play; it is not connected with the act of registration of the result by the mind of the observer.[22]

19. Ibid., p186.
20. Ibid., p41.
21. Ibid., p53.
22. Ibid., p55.

Thus far Heisenberg's position and my own appear to be very close indeed—to the point of being indistinguishable. Is not Heisenberg's 'world of *potentiae*' tantamount to the microworld, as I have conceived of it? And his realm of 'things and facts' to what I term the corporeal world? At first glance this does seem to be the case. Upon closer examination, however, a major difference comes into view. The crux of the matter is this: In the philosophy of Heisenberg we find no sharp distinction between the physical universe on a macroscopic scale and the corporeal world, properly so called. The distinction between the world of *potentiae* and the actual world must consequently be understood in terms of size or scale alone—as if the passage from potency to actuality could be effected simply by joining together a sufficient number of atoms. Consider, for instance, the following assertion: 'The ontology of materialism rested upon the illusion that the kind of existence, the direct "actuality" of the world around us, can be extrapolated into the atomic range. This extrapolation is impossible, however.'[23] One cannot but agree that 'this extrapolation is impossible'; but the question is whether physics attains to 'the direct "actuality" of the world around us' even on a macroscopic scale. My own position, in any case, is entirely clear in that regard. I maintain that the descent from actuality to potency takes place already on a macroscopic level: it takes place the moment we pass from a corporeal object X to its associated subcorporeal object SX. The fact, moreover, that SX can be described (up to a point) in the terms of classical physics does not alter the case, nor does the fact that these terms are derived somehow from ordinary experience.

My point, then, is this: The macroscopic objects of classical physics are every bit as 'potential' as are atoms and subatomic particles. I take seriously the claim of the atomic physicist to the effect that these large-scale objects are in reality composed of atoms. The fact, however, that SX is reducible to atoms does not imply that X is thus reducible; for indeed, X and SX are not situated on the same ontological plane. This is just the crucial point, to say it once more: SX exists as a potency, whereas X exists as a 'thing or fact'.

23. Ibid., p145.

Heisenberg, on the other hand, appears in effect to identify SX and X. In line with this identification, moreover, he conceives of the 'physical act of observation' performed upon a microsystem as a kind of translation of the micro- into a macrostate, such as takes place in a Geiger counter or bubble chamber. Now, according to my view, this process does not in itself take us out of the potential domain: the macrostate of a Geiger counter, for instance, conceived as a physical system, is yet situated on the physical plane. The passage, therefore, from potency to actuality is effected, not simply by the process in question, but by the fact that the Geiger counter itself is 'more' than a physical system. It is not in reality a physical process—a 'physical act of observation'—that actualizes the microstate, but the passage from SX to X (from the potential to the actual Geiger counter, if you will).

Heisenberg, for his part, maintains (as we have seen) that the transition from the 'possible' to the 'actual' is effected simply by the 'physical act of observation'. He is forced, however, to conclude that the physical act cannot explain the so-called collapse of the state vector; for this he needs to bring 'the mind of the observer' into the picture:

> The discontinuous change in the probability function takes place with the act of registration; because it is the discontinuous change of our knowledge in the instant of registration that has its image in the discontinuous change of the probability function.[24]

For my part, I find it hard to understand how the probability wave can have a 'completely objective' content if it depends upon whether the result of an experiment has been mentally 'registered' or not. If the position of a pointer, let us say, betokens a certain objective state of affairs *after* it has been 'read', why not *before*? We seem to be back in the mystical realm of Schrödinger's cat, wherein state vectors collapse at the opening of a hatch. So long, however, as one does not distinguish categorically between a physical system— be it ever so macroscopic—and a corporeal object, there is in fact no

24. Ibid., p55.

way out of this dilemma. It is in effect a theorem of quantum mechanics that physical systems do not cause state vector collapse. If one supposes, thus, that there are physical systems and psychic acts—and nothing else—then it follows that the collapse in question *must be* caused by a psychic act.

Strangely enough, however, Heisenberg himself appears not to be satisfied with the dichotomy of 'physical systems and psychic acts'. Time and again he inveighs against 'the Cartesian partition'; a 'dangerous oversimplification', he calls it.[25] And at certain moments he seems almost to recognize the corporeal domain. 'Our perceptions,' he writes in one of these non-Cartesian passages,

> are not primarily bundles of colors and sounds; what we perceive is already perceived as something, the accent being here on the word 'thing', and therefore it is doubtful whether we gain anything by taking the perceptions instead of the things as the ultimate elements of reality.[26]

In other words, what we perceive may not be just 'bundles of colors' but 'things': corporeal objects, as we say. Yet Heisenberg seems not to have realized that the Cartesian alternative—that is to say, the bifurcationist view of perception—is not only of 'doubtful' advantage, but is in fact untenable. Nor apparently did he surmise that a non-bifurcationist view of perception, pursued to its logical conclusion, could free his philosophy from its most embarrassing premise: the notion, namely, that state vector collapse is the result of 'registration'.

The philosophy of Heisenberg, then, and my own do not coincide. To be sure, there is an element of mystery in both: in one it is the enigma of state vector collapse—of Schrödinger's cat, as one might say—and in the other it is first and foremost the miracle of the corporeal domain—of this visible and tangible world—and thus of the creative Act itself.

25. Ibid., p105.
26. Ibid., p84.

IV

MATERIA
QUANTITATE SIGNATA

ONE SPEAKS OF many different physical objects: of stars and galaxies, of electromagnetic fields and of radiation, and ultimately of molecules, atoms and fundamental particles. We should remember, however, that each kind of object is conceived in relation to a corresponding observational procedure, and that consequently physical objects are not so much 'things in themselves' as they are things in relation to specific modes of empirical inquiry. As Heisenberg has pointed out, physics deals, not simply with Nature, but with what he terms 'our relations to Nature.'[1] One might put it this way: it is the experimenter himself who 'interrogates' what Heisenberg calls Nature[2]—the external reality, if you will; by the type and arrangement of his instrumentation he formulates a question, and it is of course the query that elicits the answer, the response. The diversity of physical objects—of the 'answers' which Nature gives—is prompted, thus, by the diversity of the questions which we ourselves have posed. But there is no reason to assume that this diversity of 'questions' and 'answers' carries over to the reality, to Nature as such. In contrast, therefore, to what we have termed the physical universe, the Nature of which we speak is not to be conceived as a domain or ensemble made up of physical objects. To be sure, physical objects do exist; the point, however, is that these objects partake somewhat of relativity, and are to be viewed, not as so many independent entities, but as diverse manifestations of a single and unbroken reality.

---

1. *Das Naturbild der heutigen Physik* (Hamburg: Rowohlt, 1955), p 21.
2. A term which proves to be somewhat misleading, as we shall soon see.

It should be noted, moreover, that this ontological position is not simply a matter of philosophical speculation, but is virtually forced upon us by the discoveries of physics, and most especially, by the results of quantum theory—so long, of course, as we would adhere to a realist stance. As David Bohm has pointed out, 'One is led to a new notion of unbroken wholeness which denies the classical idea of analyzability of the world into separately and independently existing parts.'[3] But clearly, the 'unbroken wholeness' to which Bohm alludes is tantamount to Heisenberg's 'Nature': to the transcendent reality, one could say, which manifests or reveals itself partially in the form of physical objects. The latter, therefore, exist— not 'by themselves'—but by virtue of the reality of which they constitute a partial expression. And whereas these manifestations are 'separate' and multiple, the reality itself remains 'unbroken.'

In light of these considerations it now appears that the so-called physical universe—with which we have been concerned in Chapters 2 and 3—does not stand alone, but points beyond itself, so to speak, to a deeper level of reality (which we have tentatively designated by the term 'Nature'). In the course of our previous reflections we have been led to distinguish between the physical and the corporeal planes; and now, it seems, a third ontological stratum has come into view—which in fact appears to be more fundamental, more basic than the two aforementioned planes. What, then, is the nature of this third domain?

❖   ❖   ❖

We have spoken of the deep reality as an 'unbroken wholeness'; but what exactly does this mean? How does one set about to conceive of an external realm that is not in fact made up of 'separate and independently existing parts'? To begin with, it behooves us to consider whether the reality in question is still subject to the spatio-temporal condition. We would find it hard, of course, to conceive of a Nature which is *not* spread out in space and time; but then, is this not perhaps what the notion of unbroken wholeness demands?

3. D. Bohm and B. Hiley, 'On the Intuitive Understanding of Nonlocality as Implied by Quantum Theory,' *Foundations of Physics*, vol. 5 (1975), p 96.

Let us examine the matter. In Newtonian days, as we know, space and time were thought to 'exist' independently of material entities. Space, in particular, was conceived as a kind of absolute receptacle into which bits of matter could somehow be introduced, and in which, thus emplaced, they could freely move about. With the advent of Einsteinian relativity, however, the picture has changed. According to the general theory, the space-time continuum carries a geometric structure which both affects and is affected by the distribution of matter it is said to contain. Space and time, therefore, prove to be inextricably connected with the material entities and events which make up the physical universe; in short, content and container have lost their independent status, and it now appears that space, time and matter—so far from being independent principles—do but constitute distinguishable aspects of one and the same reality. It follows, moreover, that the reality as such is neither space, time nor matter, nor indeed can it be contained in space or time; for it is ultimately the reality itself that in a sense 'contains' space-time—even as a cause may be said to 'contain' its effects.

Now, admittedly, physics as such is perforce incapable of recognizing its own proper objects as the effects or manifestations of a reality which in principle lies beyond its grasp. Or to put it another way: nothing on the technical plane compels the physicist to postulate such a reality. And yet it can also be said that the *bona fide* findings of physics do point in that direction. As Henry Stapp has expressed it, 'Everything we know about Nature is in accord with the idea that the fundamental process of Nature lies outside space-time... but generates events that can be located in space-time.'[4]

What, then, are some of the findings that point beyond the space-time continuum? It may suffice to mention only one—the most striking of all, I believe: to wit, Bell's interconnectedness theorem. Photons A and B, let us say, are travelling in opposite directions—at the speed of light!—and yet an observation, performed on photon A, seems instantly to affect B. What is one to make of this? Now, according to the classical ontology of 'separate and independently

4. 'Are Superluminal Connections Necessary?', *Nuovo Cimento*, vol. 40B (1977), p191.

existing parts,' one is evidently obliged to postulate some kind of superluminal transmission of influence from A to B. This problematic postulate, however, becomes superfluous the moment we recognize photons A and B as manifestations of a single underlying reality; for indeed, where there is unity or 'unbroken wholeness,' there is no need to communicate, to transmit influence through space and time. The real point of Bell's theorem, thus, or of the EPR phenomena in general, it seems, is that the twin particles involved in these phenomena are not in fact 'separate and independently existing parts.'

To be sure, they *are* 'separate' to the extent that they are contained in different regions of space-time; and admittedly, to the extent that we are able to observe either particle, they are thus contained. But then, everything points to the fact that a particle cannot be fully known by empirical means; and if it be true—as one has every right to surmise—that 'now we know in part,' then it becomes readily conceivable that a particle may transcend its manifested locus, and thus its phenomenal identity as well. In a word, there may indeed be more to the particle than meets the scientific eye— and by the same token, more than can be made to fit into a four-dimensional continuum. I should make it clear, however, that what stands at issue here is not the dimensionality of the containing manifold, but the absoluteness or relativity of containment itself. My point, thus, is not that the particle 'projects into another dimension,' but that in addition to its empirical aspect it has a nature which is not subject to 'containment' at all.

It boils down to this: Nature, though not spatio-temporal in its own right, presents itself as spatio-temporal under observation. This is to be understood, however, not in a Kantian, but in a realist sense. The point is not that spatio-temporal conditions are superimposed upon a noumenal reality by the human observer, but that the things and relations which we observe—'matter, space and time,' if you will—manifest or actualize a certain pre-existent potency, a potential which belongs to Nature as such. Once again, it is the physicist that 'poses the question,' but Nature herself that gives the response. And that response—let it be clearly understood—is indicative, not just of our human constitution, or of the arrangement of our instruments, but first and foremost of the reality itself. In the final count, what presents itself to us through the categories of space and time is none other than the reality which, in its own right, is not subject to these categories. And let me reiterate, for the sake of maximal clarity, that the conditions of space and time are not simply imposed, Kantian style, from the outside, but are potentially contained in the reality as such—even as points and lines are potentially contained in the Euclidean plane.

What, then, is a physical object? Nothing more, nor less, one is now bound to admit, than a particular manifestation of the total reality. *Qua* physical object, to be sure, it exists in space and time, and exhibits a certain phenomenal identity; and yet, in itself, it transcends these bounds, and that apparent identity. The notion of particulate multiplicity applies thus 'near the surface'—in answer to the different 'questions' we pose, or are able to pose—while 'unbroken wholeness' reigns in the unfathomable depths.

It is always possible, of course, to cling to the widespread belief that reality coincides with the space-time continuum and its multiple contents; but it appears that this habitual reduction of the real to the manifested is becoming ever more forced and precarious in light of ongoing scientific developments. Physics today militates against this constrictive *Weltanschauung*; 'Everything we know about Nature,' says Stapp, 'is in accord with the idea that the fundamental process of Nature lies outside space-time. . . .' And let us add that no single result, certainly, is more suggestive of this new idea than Bell's interconnectedness theorem. Indeed, it could well be said that Bell's theorem may be the nearest that physics can conceivably come to the formal recognition of the revised ontology which I have attempted to delineate: the view, namely, that there is not only a space-time continuum containing various entities, but also—on a more fundamental level—an as yet undifferentiated potency, which is neither in space nor in time, and about which nothing specific can be affirmed. 'Reality is non-local'; that, perhaps, is the closest we can come.

❖ ❖ ❖

But even though there is nothing in Nature—no 'thing', in other words—that we can know, the fact remains that we can and do know

Nature by way of the spatio-temporal universe. And that, after all, is what physics is about: the physicist would know 'the structure of Nature'; it is only that we are obliged to view that 'structure' indirectly, which is to say, by way of its physical manifestations.

But then, it is to be noted that even the most familiar structures of a geometric kind can only be known, likewise, through indirect means. How, for example, does one describe, or axiomatize, the structure of the Euclidean plane? As every mathematician knows, this can be done in various ways: *à la* Euclid, for instance, in terms of the properties of certain constructed figures made up of points, lines, and circles; or *à la* Felix Klein, in terms of the invariants of a continuous transformation group. The very circumstance, however, that these various characterizations are strikingly dissimilar already testifies to the fact that we are approaching the structure of the Euclidean plane by means of an auxiliary construct, a secondary structure of some kind, which presumably is more concrete and accessible. The primary structure is revealed through the secondary, one could say. In the classical approach, for example, one looks at constructed figures—but not directly at the Euclidean plane. For indeed, in the plane as such there is nothing to be seen.

Now let us substitute Nature for the Euclidean plane, and physical systems for the figures of classical geometry—and we may catch a glimpse of what physics is about. For by way of the geometric analogy one is able to understand how the structure of Nature—hidden though it be—can be manifested in the fundamental laws of physics: in the laws, namely, that apply always and everywhere to the physical systems to which they refer. A splendid example would be Maxwell's equations, which apply to every electromagnetic field—even as the theorem of Pythagoras, let us say, applies to every right triangle. The major difference, however, between Euclidean geometry and physics in its present state is that the latter does not yet dispose over a single coherent set of principles that cover the entire ground. It is as if the physicist had one set of laws for 'triangles,' and another for 'circles'— but no single law that applies to both 'circles' and 'triangles', and in principle, at least, to all other constructible figures. One might say that physics, in its present state, is conversant with 'theorems' but has not yet discovered a single set of axioms from which all the rest can in principle be derived. And this is of course the ultimate object of the physicist's quest: he is looking for a single basic law—in the form of a unified relativistic quantum field theory of some kind, presumably—that will correctly describe all conceivable physical systems. And it appears that he may indeed be approaching the realization of this goal. Such a breakthrough, in any case, would accomplish for physics what the axiomatization of the Euclidean plane has accomplished for classical geometry: it would give us a faithful representation, one might say, of the primary structure.

The objection might be raised that the laws of physics have to do—not with Nature as such—but with 'our relations to Nature,' as Heisenberg has said. The point, however, is that they have to do with both—even as the theorem of Pythagoras, for example, has to do, not only with a certain class of constructed figures, but also with the structure of the Euclidean plane. Why should the one fact exclude the other? Admittedly, Eddington has claimed that the fundamental laws of physics—including even the dimensionless constants of Nature—can be deduced *a priori* from the *modus operandi* by which the laws in question may be put to the test. From an examination of the fisherman's net, says Eddington, one can draw certain conclusions regarding the nature of the fish to be caught with that net; the fish must be larger, for example, than a certain length, and so forth. But fascinating as this philosophy of physics may be, it happens that no one has yet succeeded in this Kantian enterprise, and few physicists today, if any, would follow Eddington in his radically subjectivist claims. When all is said and done, it appears that the laws of physics speak to us not only of 'our relations to Nature,' but also, ultimately, of Nature as such.

❖   ❖   ❖

That Nature, however, proves to be highly recondite, and in fact, metaphysical. Now, to be sure, it is not easy to conceive of metaphysical realities, and it is of course impossible to picture or imagine things of that kind. However, as the physicist knows full well, we can indeed conceive of unimaginable things, and we can do so, moreover, with maximal clarity and exactitude. It is therefore by no means the case that human knowledge is restricted to the sensible order, as certain skeptics have claimed. And if it be possible to conceive of the physical (which, as we have seen, falls outside the sensible domain), then why not also of the metaphysical: of things which transcend the bounds of space and time? Thus, despite the misgivings of Western philosophers, beginning with Locke, Hume and Kant, it appears that metaphysics, thus understood, is not after all a vain or unfeasible enterprise.

As always, however, we need the support of sensible images, of an appropriate metaphor (<*metapherein*, 'to carry over') or corporeal paradigm.

What then, let us ask, is a proper metaphor for the concept of Nature at which we have arrived? What, in fact, is the paradigm which has been lurking all along in the back of our minds? It is none other, we say, than the hylomorphic or sculptural, upon which in a sense the metaphysics of Aristotle is based. This may or may not be apparent, but deserves in any case to be explained with considerable care.

Think of a piece of wood (*hyle* in Greek) or marble receiving the form (*morphe*) of Apollo or Socrates. The concrete thing—the statue—is thus in a sense composed of two factors: *hyle* plus *morphe*. It is however apparent that *morphe* has no concrete existence of its own, apart from the wood or marble in which it has been cut. But what about *hyle*? So long as we take the term in the literal sense, it has of course an existence, due to the fact that the original piece of wood has a *morphe* of its own. *Hyle* in the Aristotelian sense, on the other hand, is simply the recipient of *morphe* and nothing more. The Aristotelian *hyle* is consequently conceived as a pure substrate which stands, figuratively speaking, beneath the level of concrete existence. It is thus literally a non-entity; and yet, like the zero of mathematics, this 'nothing'—strange as it may seem—plays a crucial role. It is by virtue of this role, moreover, that we can conceive of the Aristotelian *hyle* in the first place; for in itself, as I have said, it is 'nothing'. What, then, does *hyle* do? if one may put it so. It receives *morphe*, receives content—receives being, in fact; and this it can do precisely because, in itself, it is amorphous, empty, and indeed, non-existent.

Morphe, for its part, has no concrete existence either, as noted before; it exists, so to speak, in conjunction with *hyle*—even as the form of Apollo exists in conjunction with its marmoreal support. *Morphe*, however, is not simply 'form, shape, or figure' in the more or less visual sense of these terms—we must not press the sculptural metaphor too far. The point is that the *morphe* of an existent entity is precisely its knowable aspect. In short, a thing is intelligible by virtue of its *morphe*—but existent on account of *hyle*. I do not say 'its *hyle*,' moreover, because *hyle*, strictly speaking, does not belong to the thing—any more than the ocean could be said to belong to a particular wave. *Morphe*, on the other hand, does appertain to the thing: for the *morphe* of an entity is truly its essence (<*esse*, 'to be').[5] It is what we know and can know; and thus it is the 'what' or quiddity of the thing. One must bear in mind, however, that the existent entity does not simply coincide with its quiddity: it has also a hylic aspect, which remains unintelligible—a fact of the utmost significance, to be sure.

One might note that with the revival of Aristotelian philosophy during the Scholastic era the Greek term 'morphe' came naturally to be replaced by the Latin 'forma,' and *hyle* became *materia*. And by a certain evolution, moreover, the Scholastic 'materia' became eventually transformed into the 'matter' of the Newtonian physicist—the exact meaning of which, however, is far from clear. Ontologically speaking, this remnant of the Newtonian era constitutes, in any case, a confused hybrid of *materia* and *forma* in the authentic sense. And unlike 'mass'—with which it is sometimes confounded—it has actually no rigorous role to play in the economy of scientific thought.

The closest to authentic *materia* to which Newtonian 'matter' was destined to attain, was no doubt the ill-fated ether, whose intended function it was to support the electromagnetic field. Despite its perfect homogeneity, extreme attenuation, and other 'ethereal' characteristics, however, that ether was yet conceived as a 'substance' in the contemporary sense. Authentic *materia*, on the other hand, is a

---

5. The Thomistic distinction between essence and form has no particular bearing on our present considerations and may therefore be suppressed.

thing of a very different kind. First of all, one must understand that *materia* does not occupy space—as is evident the moment one recalls that space has to do with geometric relations between existent entities. Ontologically speaking, therefore, space is posterior to *materia*; and the same, incidentally, applies to time. And yet it could also be said that space, conceived as an empty receptacle or universal container, constitutes a kind of natural symbol or cosmic image of the material substrate. Authentic *materia*, thus, so far from being characterized by extension like the Newtonian 'matter,' is on the contrary allied to the container, the pure receptacle.

A few words may be in order, at this point, regarding the once illustrious philosophy known as materialism, which purports to explain all things in terms of Newtonian 'matter' alone. Now, in the first place, it is apparent, in light of what has been said above, that corporeal existence entails necessarily *two* principles: 'It takes two to exist,' if you will. If, however, one nonetheless seeks to reduce corporeal things to a single principle, the Newtonian 'matter' turns out to be an especially poor choice. For apart from the inherent vagueness of this notion and its uselessness on a rigorous scientific plane, the concept stands yet predominantly on the side of *materia*. It represents existence denuded, so to speak, of most of its formal content, and constitutes thus a kind of near-materia or quasi-substance. The materialist, therefore, is looking towards *materia* in his quest for a single principle in terms of which everything can be understood—an unfortunate choice, seeing that *materia* is not only one hundred percent unintelligible in its own right, but lends to all things their aspect of unintelligibility, if one may put it so. The shift, therefore, from a materialist to a structuralist interpretation of physics, which came about in the wake of Einsteinian relativity, represents doubtless a turn in the right direction: from *materia* to the intelligible aspect of reality.

The fact, however, that things are intelligible by virtue of their formal aspect does not imply that they can be adequately conceived purely and simply as forms, or as structure in the physicist's sense. Thus, if materialism turns out to be untenable, so ultimately does structuralism; for indeed, in the final count, I maintain, there can be no viable ontology which does not, in one way or another, invoke the hylomorphic paradigm. The very idea of corporeal existence, one can say, is demanding of two complementary principles, which cannot but answer to the twin conceptions of *materia* and *forma*. And this explains why corresponding notions are to be found in the major ontologies, from China and India to Greece and ancient Palestine.[6]

❖   ❖   ❖

To perceive the necessity of the hylomorphic conception one needs but to reflect upon the epistemological enigma: the problem of knowledge. We have maintained that the corporeal domain is known through sense perception and the physical through the *modus operandi* of scientific observation; but what does it mean 'to know'? I have indicated that the process of knowing culminates invariably in an intellective act, but what is the nature of this act? Wherein does it consist?

As Aristotle pointed out long ago, the act of knowing consists in a certain union of the intellect with its object. But how can the intellect be joined to the external thing? Such a union, clearly, can only be conceived in terms of a third entity or common element, which object and subject can both possess, each in its own appropriate mode; and it must be this *tertium quid*, precisely, that renders the object knowable.

But only in part! For it is not, after all, the external object—lock, stock and barrel—that 'passes into the subject,' but only what I have termed the *tertium quid*. This 'third factor,' moreover, answers to

---

6. This is no doubt far more evident in the case of China, India, and Greece than it is in the case of 'ancient Palestine'. And yet it cannot be denied that the hylomorphic conception is likewise Biblical. Meister Eckhart, for one, has apprised us of this fact: 'One needs first of all to know that matter and form are not two kinds of existent entities, but two principles of created beings. That is the meaning of the words: "In the beginning God created heaven and earth"—to wit, form and matter, two principles of things.' See *Liber parabolarum Genesis*, I.28. The interested reader can find this text in the magnificent Kohlhammer edition of Meister Eckhart's works, which gives the Latin plus a German translation. See *Meister Eckhart: Die lateinischen Werke*, vol. 1 (Stuttgart: Kohlhammer, 1937–65).

the question 'What?': it is what we know. And yet it does not simply coincide with the object as such, for as just noted, the latter is perforce 'more' than the *tertium quid*.

Now the *tertium quid*, to be sure, is none other than the Aristotelian *morphe*, the form or quiddity of the existing thing. But inasmuch as the thing does not coincide with its *morphe*, one needs to postulate a second principle—an X, if you will—that distinguishes the two, or makes up the difference, so to speak. And this X—which is perforce unknowable and has no quiddity—is evidently tantamount to *materia*. One arrives thus, by way of epistemological considerations of a rather simple kind, at the basic conceptions of the hylomorphic paradigm.

It is worth pointing out that the *morphe* or *tertium quid* needs likewise to be existentiated subjectively, which is to say, on a mental plane. It needs, as it were, to be clothed in mental images and thus, in a manner of speaking, 'embodied'. The human process of knowing is complex, as we have had ample occasion to note. But yet it is consummated in a single intellective act that is perfectly simple—and for this very reason eludes analysis. And it is here—in this enigmatic act—that the cognitive union takes place: that subject and object unite.

❖   ❖   ❖

I mentioned at the outset of this ontological interlude that the idea of Nature at which we had earlier arrived relates to the hylomorphic paradigm. One would obviously like to conceive of that Nature as a *materia*; but being endowed, as we have seen, with a form of its own, it is not *materia* in an absolute sense, not *materia prima*, in Scholastic parlance. However, it does evidently constitute a *materia secunda* in relation to the spatio-temporal world in precisely the same sense in which the Euclidean plane can be termed a *materia secunda* in relation to the universe of constructed figures. As *materia*, thus, it stands 'beneath' the spatio-temporal domain in an ontological sense, as the carrier or receptacle, that is, of its formal content. And yet it owns a form which it passes on to the universe at large as a universal law or principle of order; as the least common denominator, so to speak, of the sum total of manifested forms. Nature, thus, turns out to be a *materia quantitate signata* (a *materia* 'marked by quantity'), if it be permitted to adopt this excellent Thomistic phrase.[7]

Finally, it is to be noted that the Euclidean or geometric paradigm— in terms of which I had earlier sought to explain the rationale of physics—is indeed tantamount to the hylomorphic. It constitutes in fact the form or version of the hylomorphic paradigm which relates most directly to the *modus operandi* of physics. And as such it proves to be indispensable.[8]

❖   ❖   ❖

It will be of interest, in light of these considerations, to reflect upon the time-honored distinction between 'quantities' and 'qualities' as conceived in relation to the corporeal domain. What (if any) is the ontological significance of this presumed complementarity? It is to be noted, in the first place, that inasmuch as the quantities in question pertain to the corporeal level, they must be somehow perceptible. Or to put it more precisely: It must be possible to observe or ascertain these quantities without the use of scientific instruments. Now it happens that there are two modes of quantity: 'number' in the sense of cardinality, and 'extension.' The former, clearly, is ascertained by counting, or in the case of sufficiently small ensembles, by a kind of direct perception, it would seem. Extension, on the other hand, has to do with 'large' and 'small', 'straight' and 'curved', and a

7. I am not of course claiming that the meaning which I have assigned to this phrase coincides with its original Thomistic connotation. Obviously the Angelic Doctor was not thinking of quantum field theory! And in fact, it appears that the notion of mathematical structure is inherently Platonist and somewhat foreign to the Scholastic mind. But be that as it may, the Thomistic sense of the phrase may be found in *De ente et essentia*, chap. 2.

8. One is reminded of the celebrated admonition reputedly inscribed over the portal of Plato's Academy: 'Let no one ignorant of geometry enter here.' It is doubtless no accident that geometry has occupied a central place of honor in the Pythagorean and Platonist traditions. One may presume that in its ancient or Euclidean form, this science constitutes indeed one of the major keys to 'cosmology', in the *bona fide* sense. The purport of Plato's inscription, it appears, is that no one ignorant of geometry *can* 'enter here'.

host of other geometric attributes falling within the range of human perception. The two kinds of quantities, moreover, are closely connected, and that, to be sure, is the reason why a single science—to wit, mathematics—is capable of dealing effectively with both.

The qualities, on the other hand, could be characterized precisely by the fact that they do *not* submit to mathematical description. And this is no doubt the reason why Galileo and Descartes felt obliged to ban these so-called 'secondary' attributes from the external world: The qualities had to go because they could not be made to fit into a mechanical universe, a universe that could be understood in purely mathematical terms. However, as we have already shown at great length, the qualities nonetheless exist; the redness of an apple, for instance, exists and belongs to the external object as truly as its shape. It comes down to this: An entity devoid of qualities is *ipso facto* imperceptible; for indeed, things are perceived by virtue of their qualitative content—even as countries on a map, for example, are rendered visible, not (strictly speaking) by their mathematical boundaries, but by their respective colors. And so we find that the corporeal world comprises after all both 'quantities' and 'qualities'—as most people had thought all along.

But whereas the qualities are, so to speak, ubiquitous on the corporeal plane, not a single one is to be found on the physical; for the physical domain consists, as we have seen, of things that can be described, without residue, in mathematical terms. It consists thus of mathematical structures, or of 'existentiated mathematical forms', as I have sometimes said. However, we should remind ourselves that physical objects prove ultimately to be nothing more nor less than certain 'potencies' in relation to the corporeal world. It is by no means unreasonable, therefore, to surmise that existence, properly so called, 'begins' on the corporeal plane. It could of course be objected that this is a matter of semantics, and that the epithet 'existence' may indeed be applied on the physical level as well; but then, by the same token, we are also within our rights to adopt the former stand, which is what I propose to do—in keeping with the idea that 'below' the corporeal plane one encounters *potentiae* of various kinds, and nothing more.

Now the sub-existential planes—the physical, namely, and the sub-physical *materia secunda*—are constituted, as already noted, by mathematical forms. Below the level of existence quantity alone remains. When we come to the corporeal plane, on the other hand, the qualities appear: attributes which cannot be understood or explained in quantitative terms. It is true that corporeal objects admit quantitative attributes as well. They carry in fact a certain mathematical structure which derives from the associated physical object and can be fully comprehended in physical terms.[9] And that is of course the reason why physics is possible in the first place, and why physicists have been tempted to exorcise the qualities and identify the corporeal with the physical realm. Leave out the qualities, and there remains but a single ontological domain, constituted by mathematical structure.

But as we know, the qualities refuse to be exorcised. The fact, moreover, that qualities abound in the corporeal domain but are nowhere to be found on the sub-existential planes can only mean one thing: qualities betoken essence (<*esse*, 'to be'); the essence, namely, of the corporeal entity. And that essence, let us clearly understand, is not a mathematical structure: the very fact that corporeal objects admit qualitative attributes suffices to preclude that possibility.

The corporeal domain is constituted, thus, by 'non-mathematical' essences—shocking as this may sound in our day.[10]

But let us continue. Having discerned that the qualities are indicative of essence, we need now to ask ourselves what, in that case, is the significance of quantities, and more generally, of mathematical

9. One can say, in fact, that there is a presentation-induced isomorphism between corporeal and subcorporeal quantities.

10. A few words regarding the ancient sciences, the ones which supposedly constitute 'primitive superstitions'. What one generally fails to realize is that the *bona fide* traditional sciences are primarily concerned with 'essences': the very thing which we have systematically excluded from our *Weltanschauung*. A case in point, no doubt, are the so-called five elements of the ancient cosmologies, which modern interpreters have been far too quick to identify with 'earth', 'water' and the rest, in the literal sense of these terms. For it is altogether likely that these elements are not in fact substances in the modern sense, but 'essences' of which earth, water and the rest constitute, so to speak, exemplifications. It is no accident, thus, that one of these elements was known in medieval times as the *quinta essentia* or 'fifth essence' (from whence our word 'quintessence' obviously derives). What is still more to the

forms? The answer to this question, however, has been known for a very long time; as the Scholastics used to say: *Numerus stat ex parte materiae.*[11] Quantity and mathematical structure, in other words, refer to *materia*, or more precisely, to the material aspect of things. The concrete object is made up, as we have seen, of matter and form; and this ontological polarity is reflected on the plane of manifestation. The existent object bears witness, so to speak, to the principles by which it is constituted; to both the paternal and maternal principles, if you will. And that is the reason, finally, why there are both qualities and quantities in the corporeal domain: the one indicative of essence, the other of the material substrate.

❖   ❖   ❖

In light of these considerations we are able, at last, to perceive the full magnitude of the Cartesian deviation. For it appears that in rejecting the qualities or so-called 'secondary' attributes, Galileo and Descartes have cast out what in fact is primary: the very essence of corporeal things.[12]

Now to be sure, physics deals with the quantitative aspects of cosmic manifestation; and this is obviously legitimate and informative

---

point, however, is that Hindu doctrine associates the so-called five *bhutas* and their subtle counterparts (the *tanmatras*) with the five sensible qualities; which is to say that a corporeal object is audible by virtue of *akasa*, visible by virtue of *tejas*, palpable by virtue of *vayu*, perceptible to taste by *ap*, and odiferous by *prithivi*. In a word, the so-called five elements are what makes things perceptible according to the five sensory modes; and let us add that having once understood that things are not in fact perceptible simply by virtue of their presumed 'atomic constitution', it becomes clear that there must indeed be 'elements' of this kind.

11. This dictum seems to have been interpreted somewhat as follows: Number originates by way of exemplification—in accordance with the fact that there are many horses, for example, but only a single intelligible form, to wit, the form, idea, or 'species' of a horse. The one exemplar, in other words, becomes in a sense replicated or multiplied by way of the material substrate, while the form as such remains one and indivisible, as does each individual (<*in-dividuus*) or member of the species. It appears thus that 'number' does derive, not from form as such, but indeed *ex parte materiae.*

12. To put it in Scholastic terms: they have cast out substantial forms, precisely. In the absence of substantial forms, however, the corporeal world ceases to exist.

up to a point. But one must not expect too much. For all its famed prowess, there are limits to what physics is able to comprehend or explain, and these limitations happen to be far more stringent than one is commonly inclined to suppose. As the French metaphysician René Guénon has observed:

> It can be said that quantity, regarded as constituting the substantial side of the world, is as it were its 'basic' or fundamental condition: but care must be taken not to go too far and attribute to it an importance of a higher order than is justifiable, and more particularly not to try to extract from it the explanation of this world. The foundation of a building must not be confused with its superstructure: while there is only a foundation there is still no building, although the foundation is indispensable to the building; in the same way, while there is only quantity there is still no sensible manifestation, although sensible manifestation has its root in quantity. Quantity, considered by itself, is only a necessary 'presupposition', but it explains nothing; it is indeed a base, but nothing else, and it must not be forgotten that the base is by definition that which is situated at the lowest level. . . .[13]

Now admittedly, the phrase 'explains nothing' may be excessive; but it serves just the same as a counterweight to other no less exorbitant claims, put forward by those who would 'extract the explanation of this world' from the data of physics.

Strictly speaking, the only thing about a corporeal object that one is able to understand in terms of physics are its quantitative attributes; and one is able to do so, moreover, by virtue of the fact that the attributes in question are inherited, so to speak, from the associated physical object. Beyond this physics has nothing to say. It has 'eyes' only for the physical: SX is all that it perceives, all that ever shows up on its charts. And that is no doubt the reason why physicists have been able to convince themselves (and the rest of the educated world!) that the corporeal object as such does not exist; or to put it the other way round: that X is 'nothing but' SX. It is the reason why corporeal entities are thought to be 'made of' atoms or

13. *The Reign of Quantity* (Hillsdale: Sophia Perennis, 2004), pp 19–20.

subatomic particles, and why the qualities are held to be 'merely subjective'.

Finally, it needs to be observed that this putative reduction of the corporeal to the physical has the effect of rendering ontologically incomprehensible the physical itself. One can still, of course, calculate and make quantitative predictions, but that is all. One may be able indeed to answer the question 'How much?' with incredible accuracy; but any attempt to reply to the query 'What?' leads perforce to contradiction or absurdity. This *Weltanschauung* (which in truth is not a *Weltanschauung*) does not admit of an ontology. And is this not precisely the conclusion to be drawn from the interminable 'quantum reality' debate? Moreover, it is impossible even to give an unfalsified account of the scientific methodology within the cadre of the reductionist position, for in the absence of qualities there can be no perception, and hence no measurement as well. Strictly speaking, one knows neither the corporeal nor the physical, nor have any clear conception of what physics is about. Is it any wonder, then, that physicists should have (in the words of physicist Nick Herbert) 'lost their grip on reality'?[14]

---

14. Readers of Eric Voegelin may recall his dire thesis that, due to the dominance of 'second realities' in modern times, 'the common ground of existence in reality has disappeared,' and that, as a result, 'the universe of rational discourse has collapsed.' (See 'On Debate and Existence,' reprinted in *A Public Philosophy Reader*, [New Rochelle, NY: Arlington House, 1978]). There seems to be much truth in this contention. However, Voegelin is thinking of 'second realities' of a cultural and ideological kind; it apparently did not occur to him that the paramount 'second reality'—the one that seems to underlie all the rest and has bedeviled just about everyone—is none other than the physical universe, as generally conceived. The moment one forgets that this so-called universe constitutes but a sub-existential domain—a mere potency in relation to the corporeal—one has created a monster. For indeed, the physical domain, thus 'hypostatized', becomes forthwith the prime usurper of reality, the great illusion from which a host of baneful errors spring. It is neither a small nor a harmless thing 'to lose one's grip on reality'!

V

ON WHETHER
'GOD PLAYS DICE'

ONE KNOWS THAT quantum mechanical systems are indeterminate. So far as its predictions are concerned, quantum mechanics is therefore an inherently probabilistic or statistical theory—this much is clear. What is not at all clear, on the other hand, is whether the theory is complete, that is to say, fundamental. Conceivably quantum mechanics may be dealing with certain stochastic epiphenomena generated by an underlying system of a deterministic kind. That is more or less what Einstein thought, and what those who believe in 'hidden variables' think to this day, in defiance of Copenhagenist orthodoxy. And so the celebrated debate between Einstein and Bohr goes on, and will presumably continue until the central issue has been resolved: the question, namely, whether the universe is deterministic or not.

To begin with, I would like to point out that the issue cannot in fact be resolved on a strictly scientific or 'technical' plane. The very duration of the actual Bohr-Einstein exchange alone suggests as much; for if it were simply a matter of physics, one would think that the two foremost physicists of the century could have settled the question between themselves within some reasonable span of time. But they did not settle it; and Bohr, for one, continued apparently to brood over the problem till the day of his death.[1] What is still more to the point, however, and all but clinches it, is the fact that there exist stringently deterministic theories which lead to exactly

1. The evening before he died Bohr drew a figure on his blackboard. It depicted the experimental setup of Einstein's most puzzling 'counterexample'.

the same predictions as quantum mechanics. These are the so-called hidden variables theories, originally conjectured by de Broglie and first constructed by David Bohm in 1952. To be sure, the empirical indeterminacy remains: only now it arises, supposedly, not because the universe itself is indeterministic, but because the experimentalist is in principle incapable of preparing a physical system in which the 'hidden variables' are subject to prescribed initial conditions. From a strictly scientific point of view, therefore, it appears that one has a choice in the matter. One can opt for a deterministic or for an indeterministic view of reality, for a neo-classical[2] or a quantum model—it seems to be more or less a question of taste. And tastes do differ. There are scientists of first rank who see nothing incongruous in the notion of fundamental acausality—a view epitomized by John von Neumann in the words: 'There is at present no occasion and no reason to speak of causality in Nature';[3] and again there are others, beginning with Einstein, who find it unthinkable that 'God plays dice'.

2. What does, however, need to be abandoned is the classical notion of locality; this is what John Stewart Bell has established as a theorem of quantum mechanics in 1964, and what has since been verified by certain sensitive experiments. On this basic issue modern physics has cast a definitive verdict. Unlike rigorous determinism, the classical principle of locality no longer constitutes a viable option. And one might add that on this issue Einstein had been not only at odds with Bohr, but plainly mistaken. It was however Einstein himself who blazed the trail which led eventually to the proof of nonlocality. The Einstein-Podolski-Rosen paper, in other words, has accomplished the opposite of what it was intended to do; instead of proving the incompleteness of quantum theory (an issue which is still open, to say the least), it has led to the disproof of the principle of locality, and thus to the downfall of the classical *Weltanschauung*. For indeed, the 'neo-classical' model of which we speak (i.e., the de Broglie-Bohm theory) is worlds removed from the classical picture, despite its aspect of determinism. And this may indeed account for the rather cool reception which Einstein accorded to Bohm's work.

3. To which von Neumann adds: '. . . . because no experiment indicates its presence, since the macroscopic are unsuited in principle, and the only known theory which is compatible with our experiences relative to elementary processes, quantum mechanics, contradicts it' (*Mathematical Foundations of Quantum Mechanics* [Princeton: Princeton University Press, 1955], p328). One knows today that on the last point von Neumann has severely overstated his case; his mathematical deductions do not rule out the possibility of a hidden-variables theory, as von Neumann

What, then, are we to say? If the matter cannot be settled on a scientific basis, by what means—other than 'taste'—can it be resolved at all?

Is the universe deterministic, or is it not: that is the question. There can be no doubt, of course, that a certain determinism prevails on the empirical plane. We are surrounded, after all, by phenomena—from the motion of planets to the functioning of countless man-made devices—that can be described and predicted as accurately as one may wish by the methods of classical physics. And even in the quantum domain, as we know, it happens that the evolution of physical systems is rigorously governed by the Schrödinger equation—right up to the fateful moment of state vector collapse. At this point, however, determinism (or equivalently, causality) seems to break down. And yet, even this breakdown (real or apparent, as the case may be) has in general no measurable effect on the corporeal level, where one deals perforce with statistical averages, extended over atomic ensembles of stupendous size. It is thus in reality the so-called law of large numbers that accounts for the classical determinism. And that is why von Neumann could say that 'there is at present no occasion and no reason to speak of causality in Nature.' From this perspective classical determinism reduces to a mere epiphenomenon, whereas at the fundamental level, as presently conceived, causality breaks down.

We need however to remind ourselves that there also exist corporeal phenomena (involving subcorporeal ensembles as 'macroscopic' as one could wish) in which the effects of quantum indeterminacy are not masked by statistical epiphenomena, but appear, so to speak, in plain view—which is after all the reason why these effects could be detected in the first place. That is what happens, for example, when a Geiger counter is placed in the vicinity of a radioactive source. The decay of nuclei—which according to quantum mechanics constitutes an indeterminate process—triggers then a corresponding sequence of discrete events on the corporeal level. It is of course still conceivable

had thought. It turns out, in fact, that the celebrated 'von Neumann theorem', which for long had dominated scientific thought on this question, is somewhat beside the point. See especially J.S. Bell, 'On the Impossible Pilot Wave,' *Foundations of Physics*, vol. 12 (1982), pp 989–99.

that there may be a 'hidden mechanism' within the nucleus that determines the moment of disintegration—and hence the empirical sequence—in accordance with some mathematical law, and this is in effect what the hidden-variables theory maintains. The real question, however, is whether one is obliged, on *a priori* grounds, to suppose that such a mechanism must exist.

One more observation by way of clarifying the problem: The concept of determinism does not by any means coincide with the notion of predictability. Even the staunchest advocate of determinism, after all, must surely recognize that not everything in the world can in fact be predicted. Laplace himself—that paragon of determinists—maintained only that the future of the universe could in principle be calculated, if only one knew the exact position and momentum of every particle; but it goes without saying that no scientist has ever been mad enough to suppose that such a knowledge of 'initial conditions' can in fact be elicited by scientific means, or that the requisite calculation could actually be carried out once the data had been obtained. It is true, no doubt, that a phenomenon is predictable only to the extent that it is determined; but the phenomenon may very well be determined without being predictable in a pragmatic or empirical sense—there are limits, after all, to what we humans can do.

Does 'God play dice'? That seems to be the question. And it appears that Einstein has put it well; for the wording itself suggests what by now must have become quite evident: to wit, that the issue is not in fact scientific, but incurably metaphysical.

❖   ❖   ❖

The problem, then, can only be resolved on metaphysical ground. It behooves us therefore to reflect anew on the metacosmic principles of matter and form, and to bear in mind that these twin principles are reflected in various ways within every plane of manifestation or empirical domain. In all of her aspects, as a matter of fact, Nature speaks to us, as it were, of the hylomorphic duality. A case in point, as we have found, is the distinction between quantities and qualities; for as we have seen, quantities pertain to matter whereas qualities are indicative of essence, and thus of form. Or to give a second example of particular interest: it can be shown that space corresponds to the material and time to the formal aspect of what is termed the space-time continuum. Or again: the biological complementarity of female and male (if one may venture to say so these days!). We cannot, of course, enter into a lengthy discussion of all these examples; suffice it to say that the world is full of 'hylomorphic' polarities, none of which, moreover, can be understood in depth without reference to its ontological prototype.

It will be well at this point to recall the so-called *yin-yang*, the familiar emblem of the Taoists, which could be termed an icon of the hylomorphic duality. It consists, as one knows, of a circle comprising a white and a black field, which meet in an inscribed 'S'. Within the white field, moreover, there is a small black circle, and within the black a white circle. According to the traditional interpretation, the figure represents the complementarity of *yin* and *yang*, the twin principles corresponding to the material substrate and essence (or matter and form), respectively. It is customary, however, to envision the *yin/yang* polarity, not metacosmically, but in terms of this or that cosmic manifestation; which is to say that the *yin-yang* lends itself to innumerable applications. It depicts a universal law of complementarity, not unlike the general 'complementarity principle' conceived by Niels Bohr in his later years.[4]

Always, however, *yin* stands on the side of matter, and thus represents the obscure or unintelligible aspect of the thing or phenomenon in question—which is of course the reason why it is depicted in the *yin-yang* by the color black. *Yang*, on the other hand, signifies form and thus refers to the intelligible aspect; it is therefore depicted by the color white.

But what, in particular, is the significance of the black circle within the white field, and of the white circle within the black? Clearly, what stands at issue here must be more than a complementarity in the usual sense: it must be a mutual indwelling, or *perichoresis*, as one

---

4. It was no doubt on the strength of a profound intuition that Bohr selected the *yin-yang* as his heraldic emblem.

could say. And as we shall presently see, herein lies the key to the problem of determinism.

Let us get back to physics. Physics is concerned, obviously, with certain mathematical determinations, which as such stand clearly on the side of *yang*. But then, within this context, what does *yin* signify? What else could it be than a certain corresponding indeterminacy. In light of the *yin-yang* one can therefore conclude that 'in the midst' of determination, indeterminacy must somehow appear. But how? Prior to 1925, who could have envisaged such an eventuality? But this is precisely what has come to pass. The most accurate of physical theories ever conceived by man has yielded—as a mathematical theorem, no less!—a so-called principle of uncertainty. Within the white field a black circle has unexpectedly come into view. A very small circle, as it turns out, whose radius, if you will, is of the order of Planck's constant.

My point has now become evident: The interplay of determination and indeterminacy as conceived by quantum theory, so far from being in any way unreasonable, happens to be exactly what the *yin-yang* doctrine demands. Quantum indeterminacy, so far from being an inexplicable aberration, turns out quite simply to be the *yin*-side of the coin. Contrary to our classical expectations it thus appears that determination and indeterminacy are not in reality opposed or mutually exclusive, but seem in fact to imply each other in a certain high and marvelous sense. The conception of a perfectly deterministic universe proves, therefore, to be chimerical; which is to say that God does in a sense 'play dice'—distasteful as this *lila* may be to the Cartesian rationalist.[5]

❖   ❖   ❖

'The universe,' someone has said, 'is a fabric woven of necessity and freedom, of mathematical rigor and musical play; every phenomenon participates in these two principles.'

---

5. Inasmuch, however, as God does not act 'in time', it can also be said that God does not 'play dice'. I shall return to this aspect of the question in the next chapter.

It appears, however, that of these two principles, the second was roundly forgotten during the Newtonian era. With the waning of the Middle Ages it seems that a pronounced bias in favor of 'law' began to manifest itself. Not only did men come to believe more keenly in the existence of universal laws, but they began to imagine that all movement and transformation pertaining to the corporeal domain could be rigorously accounted for on the strength of law alone. And this supposedly all-comprehending law came soon enough to be conceived in strictly mathematical and indeed mechanical terms, in accordance with what may be called the clock-work paradigm. There has been much debate regarding the causes of this intellectual evolution, which may have ranged from a decadent Scholasticism gone awry to the actual construction of mechanical clocks;[6] but what especially concerns us is the fact that the movement culminated in the Cartesian philosophy. It was Descartes, after all, who formulated the doctrine of mechanical determinism in its full-blown format, and thereby laid the theoretical foundations upon which the edifice of Newtonian physics was subsequently based; and so it was by way of the Cartesian legacy that the specter of a clockwork universe eventually imposed itself upon Western civilization. In retrospect one can say that from the so-called Enlightenment right up to the time of Max Planck, that *Weltanschauung* has reigned supreme; and even today, in the face of quantum indeterminacy, it remains with us as a formidable influence. What else, after all, stands at issue in the Bohr-Einstein controversy than a certain residual Cartesianism? Why else should a physicist be so vehemently opposed to the idea that 'God plays dice'? There are those, to be sure, who would justify their opposition to the idea of indeterminism by one argument or another, and Stanley Jaki, for one, has gone so far as to perceive in the Copenhagenist

6. Already in the fourteenth century one encounters a marked predilection for certain kinds of astronomical clocks that were no doubt strongly suggestive of the clockwork paradigm. As one historian has described the scene: 'No European community felt able to hold up its head unless in its midst the planets wheeled in cycles and epicycles, while angels trumpeted, cocks crew, and apostles, kings and prophets marched and counter-marched at the booming of the hours' (Lynn White, *Medieval Technology and Social Change* [Oxford: Oxford University Press, 1962], p124).

stance 'a radical inconsistency', resulting supposedly from 'the radical rejection by that philosophy of any question about being.'[7] But whereas it may be true that physicists subscribing to the Copenhagen school of thought have as a rule paid scant attention to ontology, I maintain that only a lopsided ontology—one that conceives the world to be made of *yang* alone—could lead its votaries to believe in a Cartesian-style determinism, or to perceive a 'radical inconsistency' in its denial.

Admittedly, an event is intelligible to the extent that it exemplifies a law, a formal principle of some kind; and by the same token, whatever does not fall thus under the aegis of a law is *ipso facto* unintelligible. But nothing says that the unintelligible cannot occur, distressing as the idea may be to the rationalist. One has no *a priori* grounds to suppose, for example, that the disintegration of a radioactive nucleus must in principle be subject to a deterministic law—no matter what quantum theory may have to say on the question.

Getting back to Descartes, it is of interest to point out that along with the tenet of bifurcation and the ill-fated *res extensae*, the French savant has introduced a third notion of major importance: to wit, his so-called analytic geometry. The basic idea—familiar to any student of mathematics—consists in the supposition that the mathematical continuum, be it a line, a plane, or a space of higher dimension, can be 'coordinatized', and thus conceived in effect as an infinite point set. One knows today that this step is not quite as unproblematic as one had once thought, and some contemporary mathematicians of repute have gone so far as to reject the idea of infinite sets. What one generally fails to see, however, is that the so-called coordinatization of the continuum has destroyed—or better said, obscured—the distinction between 'potency' and 'act' within the mathematical domain. According to the pre-Cartesian conception, as we have noted before, there are no points on a line or in the plane—until, that is, these points have been somehow constructed. Points, in other words, represent determinations, whereas the continuum as such constitutes a kind of material substrate or

7. 'From Scientific Cosmology to a Creative Universe,' in *The Intellectuals Speak Out About God*, edited by Roy A. Varghese (Chicago: Regnery Gateway, 1984), p71.

'potency'—which is the reason, let us recall, why the example of geometry could be used as an ontological metaphor. One can say that the continuum represents the material principle in the quantitative domain, the black half, if you will, of the circle. But this is precisely what the French rationalist was bound and determined to extirpate, be it in the external universe or in the domain of its mathematical representation. In either realm the 'black' had to go. With the introduction of what is to this day referred to as a Cartesian coordinate system, the Cartesian opus is complete.

❖ ❖ ❖

But nonetheless the 'black' remains. And what is more, by virtue of that marvelous *perichoresis* of which the *yin-yang* tells, it actually combines with the 'white'. At the heart of all things there is to be found a certain *coincidentia oppositorum*; and herein, as I have said, lies the key to our problem: the enigma of indeterminism. The astounding fact is that freedom and necessity can coexist; the one does not exclude or cancel the other, as one commonly thinks. Thus, in the midst of necessity, freedom can exist; and not simply as a foreign element—not in reality as a black spot in a white field—but as something intimately linked to necessity as to its counterpart. In a word, there is a certain union of freedom and necessity, which moreover presents itself in countless modes. All art, for instance, is based upon such a synthesis. In a musical composition, for example, tonality and meter stand on the side of 'necessity'; it is within this framework, this 'law', that the composition is constrained to unfold. What is unfolded, on the other hand—the melodic content, so to say—is by no means determined or necessitated by that tonality and meter. A true work of art, as we know, displays invariably a marvelous freedom, which in fact is only heightened by the stringency of the prescribed form. It is precisely *within* a given law or prescribed canon that a genuine freedom of artistic expression can be achieved. As Goethe has said, *In der Beschränkung zeigt sich der Meister* (In delimitation the artist shows himself).

Having spoken of 'freedom' in the context of art, I should not fail to point out that this term does not by any means betoken the arbitrary or the accidental. Freedom of expression does of course presuppose a certain 'indeterminacy' or leeway in the prescribed bounds; but the passage from potency to act is not in reality effected by the casting of a die. It is accomplished evidently by the artist, the intelligent agent who expresses or reveals himself *in der Beschränkung*, that is to say, in a subjection to certain bounds.

Let us attempt to understand this interplay—this 'dialectic of freedom and necessity'—as clearly as we can; for it happens that much rides upon this question. The creative act, I say, consists in the free imposition of a certain bound, a certain determination. That new determination, however, is something quite different from the original or pre-assigned bounds. One needs thus to distinguish clearly between bounds 'from below', which are somehow given, and bounds 'from above', which are freely imposed. It is to be noted, moreover, that the second *can* in fact be imposed precisely because the first have left a certain leeway or 'indeterminacy'. It is by virtue of such an indeterminacy that tonality and meter, for example, can serve as a canon for musical composition.

But there is more to be said. It happens that there is a certain harmony or kinship between the two kinds of bounds; for not only is the artist heedful not to transgress the prescribed canon, but as one knows, he carefully selects this 'constraint' with an eye to the artistic idea he wishes to express.

❖   ❖   ❖

Before leaving the subject of art it behooves us to observe that art in general alerts us to a metaphysical fact of the utmost importance. For the example of art obliges us to recognize that the hylomorphic paradigm, as we have conceived of it up till now, is incomplete and insufficient. We have all along been looking at only half the picture: the lower half, as one could in fact say.

Let us then get back to our hylomorphic starting point and ask ourselves: How does a piece of uncut marble acquire the form of Socrates? The first thing to note is that the answer to this question cannot be framed in terms of *materia* and *forma* alone; that is to say, one requires, once again, a *tertium quid*, which now, however, must answer to the idea of an agent or active principle: the artist or sculptor, namely, who bestows the form. And that form, moreover, must somehow pre-exist as an archetype, or as 'the art in the artist', to use another Scholastic phrase. One sees, thus, that the hylomorphic paradigm, in its full-blown format, must entail, not two, but four ingredients—corresponding precisely to the so-called material, formal, efficient and paradigmatic 'causes' of Aristotle.

On the other hand, it is not necessarily inadmissible to neglect the distinction between the efficient and the paradigmatic or final cause—between the 'artist' and the 'art in the artist'—and thereby to combine the two in a single active principle. What we can by no means afford to leave out of account, however, is the idea of an agent or active principle as such. We need, thus, to recover the distinction between *natura naturata* and *natura naturans*: the 'natured' and the 'naturing'—to put it, once again, in Scholastic terms.

But as we know, the idea of a metacosmic agent—of a *natura naturans*—has fallen into academic disfavor; and so, too, the word 'nature' has lost its higher connotation, and has come to refer exclusively to this or that aspect of *natura naturata*. After all, having cast out the notion of 'forms', there is no further need for a 'form-bestowing' agent. The word is out that 'evolution' takes care of the genesis problem: from the universe as a whole down to a species of microbes, everything simply 'evolves'. Now, things do no doubt evolve; but only after they exist, after they have received a form or nature that *can* 'roll out or unfold'. And so, in the final count, the fact remains that *natura naturata* does presuppose *natura naturans*: the natural presupposes the supernatural—distasteful as this truth may be to some. And as to the Scholastic term 'natura naturans', it constitutes of course a *nomen Dei*: it refers to God, conceived as the 'giver of forms'.

❖ ❖ ❖

We are now in a position, finally, to consider 'the union of freedom and necessity' in the context of physics, which is after all our primary concern. Where, first of all, does that union take place? It takes place, I say, in the phenomenon of quantum indeterminacy.

Let S be a physical system and X an observable of S, and let us suppose that S is not in an eigenstate of X. The value obtained in a measurement of X is then indeterminate. The measurement can in principle yield any value belonging to the spectrum of X; there is, I maintain, no law that determines what the outcome shall be. On the other hand, the state vector of S does nonetheless determine an associated probability distribution, which means that the transition from system to empirical outcome is not after all indeterminate in an unmitigated sense; for if the process be conceived, let us say, in terms of the casting of a die, the latter must indeed be 'weighted' according to a prescribed law.

Certainly the quantum mechanical probability distribution associated with a given observable does not determine the result of a measurement. But yet it must have as much to do with that result as the weights of a loaded die have to do with the outcome of a toss; for indeed, on a statistical level the two cases are indistinguishable. The question, however, is this: In the case of an actual die, the influence of the weights is effected by way of a temporal process, which moreover is strictly deterministic. The motion of a die, after all, is determined by the equations of classical mechanics, which is to say that indeterminacy enters the picture by virtue of our inability to control the initial conditions with a sufficient degree of accuracy. The case is consequently analogous to that of hidden variables. But is it possible to conceive of quantum indeterminacy along these lines? Is it in fact legitimate to suppose that the outcome of a measurement is in reality the result of some temporal process, *be it deterministic or not?*

It appears, in light of quantum theory, that this question is to be answered in the negative. For the collapse of the state vector associated with a determination of X presents itself as a discontinuity, and thus as an instantaneous event, if such an expression be allowed. And unlike the discontinuities one encounters in the classical domain, this quantum mechanical discontinuity does not arise from an underlying continuity by way of approximation, but proves to be irreducible in principle to any continuous temporal process. Now admittedly, *Natura non facit saltus*: Nature does not 'jump'; but it needs be understood that this dictum applies to 'nature' in the ordinary sense: to *natura naturata* as distinguished from *natura naturans*. Meanwhile, strangely enough, it is in fact the characteristic of *natura naturans* to act, not by way of a temporal process, but 'instantaneously', as it were. Continuity, one can say, is indicative of the material substrate, whereas discontinuity is indeed the hallmark of the creative act.

Our point has now become evident: The significance of the quantum mechanical discontinuity—the significance of state vector collapse—lies in the fact that it betokens an action of *natura naturans*. There is a certain transition from potency to manifestation—from the physical to the corporeal plane—and such a transition can only be effected by the creative or 'form-bestowing' principle, which is *natura naturans*. But since the action of *natura naturans* is perforce 'instantaneous' (a matter to which I shall recur in the next chapter), it turns out that there is in reality no temporal process—no actual 'roll of the die'—which determines or selects the measured value of X from the spectrum of possible outcomes. This determination derives, so to speak, 'from on high,' and interrupts the normal course of events, that is to say, the Schrödinger evolution of the given physical system.

The phenomenon of quantum indeterminacy can now be understood by analogy to the phenomenon of artistic production.[8] Once again there are two kinds of bounds: the bounds 'from below', first of all, consisting in the probability weights of the state vector; and the bounds 'from above', that is to say, the measured values of the given observable, as revealed in the final state of the corporeal instrument. And these two kinds of bounds, obviously, are quite different: so much so that they pertain in fact to distinct ontological planes.[9] The apparent freedom, moreover, in the imposition of the

8. The scenario of quantum mechanics turns out thus to be rigorously analogous to the example of art. The reason for this analogy, moreover, is suggested by the Scholastic dictum: 'Art imitates Nature'—Nature, that is, in the sense of *natura naturans*.

9. In light of our considerations in Chapter 4, it appears that the determinations by which *potentiae* are actualized on the corporeal plane must entail qualitative as well as quantitative bounds. Quantities alone, as I have noted repeatedly, do not yet constitute a corporeal entity.

final determinations, obviously presupposes, once again, a corresponding indeterminacy on the part of the pre-assigned bounds.

What greatly puzzles us, on the other hand, is the fact that the results of measurement fulfill (by their relative frequencies) the demands of the pre-assigned probability weights—as if by a miracle—in a kind of spontaneous 'dance' that defies causal analysis. The metaphysical significance of this enigma has however become clear. The phenomenon can be understood by analogy to art; what confronts us here is a *bona fide* union of freedom and necessity: of 'mathematical rigor and musical play.'

VI

VERTICAL CAUSALITY

THE REFLECTIONS OF THE PRECEDING CHAPTER have brought to light a major truth: contrary to the presuppositions of modern scientific thought, the observable universe is not ultimately intelligible on the basis of natural causality; to put it in Scholastic terms: *natura naturata* presupposes *natura naturans*. The natural or 'natured' world presupposes a creative or 'form-bestowing' agency not simply in the sense of a first cause that brought the universe into existence, but as a transcendent principle of causality that is operative here and now. This is the conclusion at which we have arrived prompted by the phenomenon of state vector collapse; so far from constituting merely a conundrum of quantum theory, the significance of that 'collapse' proves to be in the first place metaphysical. What stands at issue is the validity of naturalism, the postulated hegemony of natural causation. It turns out that the observable universe does not after all answer to the conception of a closed system; not only is there a Metacosm, but one is finally forced to conclude that the spatio-temporal universe neither exists nor functions on its own.

It has been said, often enough, that quantum mechanics has invalidated the postulate of determinism, the notion that the state of the universe at any initial moment of time determines its future states. In place of a rigid determinism the new physics has supposedly arrived at the conception of a partly random universe, in which there is scope for what, by default of intelligible lawfulness, is termed 'chance'. The predictability of the Newtonian universe, we are told, is inherently statistical, and applies to macroscopic ensembles involving a vast number of fundamental particles, whereas on the level of the particles themselves the element of chance comes into play, and such laws as still apply do not suffice to determine the outcome of natural processes. Yet, while it is certainly true that the classical determinism has been overthrown, it is nonetheless misleading to speak of 'chance' in reference to the microworld. As I have pointed out before, the collapse of a state vector—which singles out one particular eigenstate from an ensemble of eigenstates[1]—is not actually comparable to the toss of a die; for whereas the latter constitutes a temporal process, indeterminate though it may be, the collapse of a state vector cannot be thus conceived. Let it be said apodictically that state vector collapse is not the result of a temporal process, be it deterministic, random, or stochastic.[2] A higher order of causality enters the picture, which needs to be distinguished categorically from temporal causality in any of its modes; that so-called 'collapse', it turns out, can be attributed no more to chance than to determinism, but actually entails a kind of causality which, strange to say, is 'not of this world'.

❖   ❖   ❖

Modern science, by the nature of its *modus operandi*, is incapable of grasping that kind of causality; it is unable, in fact, even to recognize that the phenomenon of state vector collapse cannot be dealt with by the means at its disposal—which accounts for the unending efforts on the part of physicists to do just that. It matters not whether time is viewed *à la* Newton as a linear continuum or in Einsteinian terms as implicit in the space-time continuum: in either case a causality transcending the temporal domain is scientifically inconceivable. Yet it is the contention of traditional metaphysics that the primary causation does in fact transcend the bounds of time. To understand what this entails we need first of all to relinquish the

1. In the general case, in which the underlying Hilbert space is infinite dimensional, this way of putting the matter is not accurate; the difference, however, is not relevant to the point at issue.

2. A stochastic process is one in which randomness and determinism both come into play, as in the case of Brownian motion, in which the trajectory of a classical particle experiences random breaks due to chance collisions with nearby particles of some kind.

notion that the universe 'exists in time'—as if time itself could transcend the universe. It is chimerical to suppose that time—at least as
we understand the term—has any reality apart from temporal process, that is to say, apart from the motions and transformations of
the natural world. According to ancient belief, time came into existence with the celestial bodies which measure its flow by their prescribed revolutions; the connection, thus, between time and the
celestial clock which measures or 'metes out' durations is such as to
render the two inseparable.

If it be granted, therefore, that the universe is not self-caused, it
follows that the creative act by which it was brought into existence
was indeed supra-temporal; as St. Augustine has put it: 'Beyond all
doubt, the world was not made *in* time, but *with* time.'[3] Yet, even so,
we tend to think of the creative act as something that took place
long ago, which is to say that we think of it all the same as a temporal event. It appears that our mind is more or less bound to think in
spatio-temporal terms even when the intentional object precludes
spatio-temporal bounds. Every mathematician, to cite a prime
example, is cognizant of the fact that spatio-temporal phantasmata
accompany even his most abstract and sophisticated reflections,
and has in fact learned the art of using images of this kind as a
means of 'seeing' the mathematical objects in question. The fact is
that images can be viewed as signs pointing beyond themselves to a
transcendent object or reality which they somehow represent. The
very possibility of metaphysical thought, in particular, hinges upon
this principle; what is required are symbolic representations of
metaphysical truths: metaphysical icons, if you will, which can be
received by our mental faculties and grasped by the intellect. Contrary to a popular misconception, the human intellect does not
operate by way of reasoning, but precisely through an act of vision
mediated by an image, an iconic representation of some kind.

Getting back to the misbegotten idea that the universe was created 'long ago', the question arises whether a suitable symbolism can
be found in terms of which the supra-temporal nature of creation
can be understood. I propose to approach this question stepwise in

---

3. *De civitate Dei*, 11.6.

terms of three observations; the first point is this: The natural way to depict a metacosm iconographically is by means of a higher dimension. Restricting ourselves to representations in the plane, this entails that the 4-dimensional space-time will have to be depicted by a 1-dimensional figure, and thus by a line or curve. If we think of the three suppressed dimensions as spatial, the resultant line or curve will then represent the empirical universe as a temporal process, or if you will, as Time itself. And this brings us to the second observation: Since every imagined point in time has a before and after, our choice is between a line open at both ends or a simple closed curve. Now, the first possibility is unacceptable iconographically because it is not actually constructible; this leaves the circle — the simplest bounded curve—as the prime candidate.[4] The third point pertains to the creative Act itself, which is now conceived beyond time, and thus metacosmically. What needs to be recognized is that this Act is perfectly simple: it is undivided, and in fact indivisible. It must therefore be represented iconographically as a point. That point, however, must be unique, set apart from all other points by some mark of distinction, of preeminence. With this third stipulation, however, the defining elements of an iconographic representation have come into view: the icon must consist of a circle together with its center. I will mention in passing that this representation applies not only to the universe as a whole—to the macrocosm—but equally to every authentic being contained in that universe, and above all to man, the microcosm *par excellence*. One should add that the icon which we have characterized, so to speak, in its archetypal simplicity, admits of countless elaborations, each adapted to a particular application or domain, and was known in one form or another to every traditional civilization.

By way of contrast, it is to be noted that the modern West constitutes in fact the first civilization that does *not* view the cosmos through the lenses, so to speak, of this icon. Our science, clearly, has no use for a Metacosm, and is committed to viewing the empirical universe as a closed system that can be understood, in principle and

4. Concerning the circular or 'cyclic' view of Time I refer to Robert Bolton, *The Order of the Ages* (Hillsdale, NY: Sophia Perennis, 2001), chap. 5.

without residue, in terms of natural causality. We have done away with the notion of transcendence, and have reduced the idea of causality to the level of temporal process. One could say, somewhat hyperbolically, that Time has become the new god and Evolution the new religion.

But let us get back to our icon, which implicates a very different *Weltanschauung*. The first thing to observe is that the creative Act has lost its status of 'long ago': for not only is the Center outside the circle of Time, but it is in fact equidistant to all the points on the periphery. Every 'here' and every 'now' participates equally in that transcendent Act, which in its own right is one and indivisible. One is able thus to understand that even though that Center is nowhere in space or time, it is yet in a sense ubiquitous: in the words of Dante, it is 'where every *where* and every *when* are focused.'[5] It likewise follows that creation is not in reality sequential; as we read in Ecclesiasticus: 'He that liveth in eternity created all things at once.' (Eccl. 18:1) 'There is an end, then,' says Philo Judaeus, 'to the idea that the universe came into being "in six days"'; and Meister Eckhart puts it more plainly still: 'God makes the world and all things in this present *now*,' declares the German master. Multiplicity, it turns out, pertains not to the creative Act, but to the created order: in terms of our icon, it pertains not to the center, but to the circumference.

It needs further to be noted that our icon comprises, not two, but three basic elements: in addition to a center and a circumference, it entails radii which connect the center to points on the circumference. This too has its metaphysical significance, its ontological interpretation; as Shabistari, the Persian Sufi, has succinctly put it: 'From the point comes a line, then a circle.' The radii represent what may be termed the 'vertical' direction, which has to do, not with spatio-temporal, but with ontological relations. Everything within space and time exists by virtue of that vertical dimension; as Shabistari has it, the line precedes the circle—not temporally, to be sure, but ontologically. It is a modern superstition that things exist by themselves, or on account of other 'things': the eclipse of verticality, ranging from disregard to actual denial, constitutes indeed the decisive step

5. *Paradiso*, 29.12.

which takes us into the modern world. Meanwhile it remains true, now as before, that the human mind has access to the vertical dimension, and that we are in fact cognizant of that officially 'inexistent' dimension not only in our moral, aesthetic and religious sensibilities, but in daily life. Even the simplest act of sense perception is consummated by the intellect, and thus transcends the bounds of space and time.[6] The external object too, moreover, transcends its spatio-temporal locus by virtue of its substantial form, failing which it could not be known. One sees that in a universe bereft of verticality objective knowledge has become unthinkable; and as a matter of fact, after centuries of futile endeavor to explain how we are able to know the external world, Western philosophers have apparently become persuaded that indeed we do *not* know. As I have argued elsewhere,[7] the modern world-view carries within itself the seeds of postmodernism; once it has been forgotten that the circle derives from the center by way of a line, the die has been cast.

❖   ❖   ❖

It emerges that there exists a primary causality which acts, not in some distant past, but in every *here* and *now* without exception. All things existing in space and time are not only brought into being, but held in existence, by this primary causation which derives from a single and indivisible Act. Unlike the kinds of causality with which modern science is concerned—which may be termed temporal or natural causation—this primary causality does not act from past to future by way of a temporal process, but acts directly, unmediated by any chain of temporal events. The question arises now whether this 'temporally unmediated' mode of action—which we shall designate by the adjective 'vertical'—is the exclusive prerogative of primary causation, or whether perhaps there exist secondary modes of vertical causality. In answer to this question it can be said that the

---

6. See 'The Enigma of Visual Perception' in my *Science & Myth* (Philos-Sophia Initiative, 2023).

7. 'Science and the Restoration of Culture,' *Modern Age*, vol. 43, no. 1 (2001).

causation effected by an intelligent agent is perforce vertical.[8] Take the case of art in the primitive sense of human making: the entire process hinges in fact upon such a vertical act. What stands at issue in authentic art is a veritable *imitatio Dei*: the human artist 'participates' to some degree in the creative prowess of the First Cause: 'All things were made my Him, and without Him not anything was made' (John 1:3). But does this mean that all production—even the shoddiest artifact—is to be ascribed indiscriminately to God Himself? Assuredly not. It is interesting to note, in this connection, that according to the punctuation which became generally accepted in post-medieval times, John 1:3 actually reads: 'All things were made by Him, and without Him was not anything made *that was made.*' We may take it that the *quod factum est* refers to what is *truly* made, and therefore to what truly *is*. The difference, Scholastically speaking, lies in the presence or absence of *form*: it is form—a transcendent element!—that bestows being. Now, the bestowal of form constitutes an incurably vertical act of causation.

There is a crucial difference, however, between forms bestowed by primary causation and forms imposed by secondary acts of vertical causality. It is the prerogative of the First Artificer to bestow *substantial* form: the forms that bring into existence the primary substances which constitute the corporeal domain, and upon which all secondary modes of production are constrained to operate: the marble, for instance, upon which the sculptor acts. The forms imposed by human art are of a different kind: they are forms that give being, not to substance, but to an artifact. One sees that despite the reality of 'participation', the *imitatio* is yet worlds removed from the primary Act itself. And yet the fact remains that the bestowal of form—be it 'substantial' or not—hinges upon a vertical act, as I have said.

Having pointed out the ubiquity of primary causation, and having noted that there exist also secondary modes of vertical causality, it needs to be stated that, even so, temporal modes of causality, too, exist and have a role to play. The primary action does not obliterate the temporal modes: on the contrary, it brings these secondary modes of

8. Ibid., pp194–98.

causality into existence and renders them operative. Temporal causation, however, is limited in its scope; one can say that it is able to effect change, effect transformations of various kinds, but cannot give rise to something new: authentic 'making', as we have seen, is the prerogative of vertical causality. To be precise: A truly productive cause is either the primary causation itself, or else it is the free act of an intelligent agent who 'participates' in the primary causality; nothing, on the other hand, is ever truly 'made' by natural causes.

It should be pointed out that these reflections relate intimately to a mathematical theorem, discovered by William Dembski, which forms the basis for what is currently known as ID theory (the initials standing for 'intelligent design').[9] What Dembski has shown is that ID can be recognized by means of a criterion, a signature which cannot be duplicated by natural causes. The theory can be formulated in information-theoretic terms and hinges upon a concept of complex specified information or CSI. The pivotal result is a conservation law for CSI which affirms that the amount of CSI in a closed system cannot be increased by any natural process, be it deterministic, random, or stochastic.[10] This means, according to our analysis, that only vertical causation is able to generate CSI. I will mention in passing that this result poses a formidable problem for Darwinist biology, since it demonstrates that the Darwinian mechanism—which in fact constitutes a stochastic process—could not have generated the immense amounts of CSI exemplified in living organisms. What presently concerns us, however, is something far more general: If it be the case that to 'make' is to produce CSI, and if it be further true that all vertical causation derives from the First Cause—either directly or through 'participation'—then it does follow, even on rigorously mathematical grounds, that 'All things were made by Him.'

Having distinguished vertical from temporal modes of causality (which could equally well be termed 'horizontal') it behooves us to

9. A summary account of ID theory may be found in *Ancient Wisdom and Modern Misconceptions*, op. cit., chap. 9.

10. Strictly speaking, deterministic and random processes are limiting cases of a stochastic process.

note that the two kinds of causation coexist without any confusion of effects: even as horizontal causation cannot produce the effects of vertical causality, so also can it be said that vertical causation does not produce effects proper to horizontal causality. I find it remarkable that this too can be understood in terms of our geometric symbolism: as every student of mechanics knows very well, a vertical force vector will not effect a horizontal acceleration, nor will a horizontal force produce a vertical acceleration. Now, it may seem that this apparent inability on the part of vertical causation to produce 'horizontal' effects is incompatible with the tenet of primacy; but this is not in fact the case. My point is that vertical causation effects *ontological* change, which in turn can affect the temporal course of events without altering the operation of horizontal causality: when a thing is changed inwardly, its outward behavior will change accordingly. The same principles of temporal causality are operative before and after the ontological alteration, and yet the resultant process exhibits a corresponding change. Meanwhile neither the ontological nor the resultant 'trajectoral' alteration is the effect of a temporal process, which is to say that both present themselves as an irreducible discontinuity. Despite the fact, therefore, that vertical causation does not directly effect temporal change, it is able to alter the course of events without any suspension of temporal causality.

❖   ❖   ❖

Getting back to the subject of quantum physics, it behooves us now to consider once more the categorial distinction between a corporeal object X and its associated physical object SX. Having brought into play the Scholastic notion of substantial form, we should point out, first of all, that what we normally take to be corporeal objects in the inorganic domain are rarely defined by a single substantial form. What confronts us in these cases is not a single substance, but an aggregate consisting of many substances, what the Scholastics termed a mixture. However, basic as this distinction may be, one sees that it bears no particular relevance to the question at hand, which is to say that we may suppose, without any real loss of generality, that X is a substance.

What, then, is the relation between X and SX? We may put it this way: What presents itself to the eye of the physicist as an aggregate SX of quantum particles is in fact a corporeal object X by virtue of a substantial form, and what accounts for the difference is indeed an act of primary causation. The quantum particles which make up SX exist as intentional objects of physics but not as components of X: as parts of X these putative particles are no longer physical, and can no longer, strictly speaking, be conceived as particles. As part of a corporeal entity, they participate in the being of that entity, that is to say, in its substantial form. We have concluded in Chapter 4 that physical particles lack essence and therefore lack being: that is why Heisenberg has situated these so-called particles 'just in the middle between possibility and reality'; it is the reason why Erwin Schrödinger concludes that

> We have been compelled to dismiss the idea that such a particle is an individual entity which in principle retains its 'sameness' forever. Quite to the contrary, we are now obliged to assert that the ultimate constituents of matter have no 'sameness' at all.[11]

They have no 'sameness', let us add, because they have no essence, no quiddity, no substantial form of their own. As has been sufficiently explained in the preceding chapters, they are not in truth 'things' but belong to the ontological category of *potentiae*. Now, the act to which they are in potency, I say, is none other than incorporation into a corporeal entity. It follows that once incorporated, they are no longer *potentiae*, and are therefore quantum particles no more. It needs however to be understood that they continue to exist as intentional objects of physics, and that the quantum-mechanical representation of SX retains validity from a physical point of view, subject however to the following proviso: it is necessary to suppose that the range of superpositions in SX is limited by the corporeal nature of X. Let us recall that this is precisely the 'de-superposition principle' which resolves the Schrödinger paradox: it is the reason why cricket balls do not bilocate, and why cats cannot be both dead and alive. It appears that the subcorporeal status of SX does have

---

11. *Science and Humanism* (Cambridge: Cambridge University Press, 1951), p17.

quantum-mechanical implications, a fact which can now be seen as an effect of vertical causality.

It follows from these considerations that corporeal entities are not in fact 'made of particles' as almost everyone staunchly believes. It matters not whether we conceive of these 'constituent particles' classically or quantum-mechanically: the notion proves to be chimerical in either case; for as has been noted, once incorporated, these putative particles are particles no more. Having entered into the composition of a corporeal being, they have become metamorphosed into something that no longer answers to the conception of a particle: they have become transformed into *bona fide* parts of an ontological whole. As such they have no separate existence, but derive their existence from the whole of which they are a part. Contrary to current belief, it is not the constituent particles that bestow existence upon a corporeal entity, but it is the latter, rather, that bestows existence upon its constituent particles by elevating them from their status of *potentiae* to that of actual parts.

It should be noted that these reflections shed light on the phenomenon of indeterminacy, which physicists look upon, more often than not, as a kind of anomaly or flaw. As if it were not bad enough that God 'plays dice', the quantum facts preclude in addition that the fundamental particles, upon which physicists had set their hope, can even qualify as 'things'. What the physics community has so far failed to grasp is that these seeming deficiencies are indeed precisely what is required in order that the particles in question may enter into the constitution of corporeal entities. To put it in a nutshell: If quantum particles did not partake of indeterminacy, they could not receive determination as *bona fide* parts of a corporeal whole. The physicists have it backwards: In reality it is not the function of particles to *bestow* being upon an aggregate, but rather to *receive* being from a substantial form.

❖   ❖   ❖

We are now at last in a position to understand the phenomenon of state vector collapse from a traditional metaphysical point of view. Early in this book it became clear that the categorical distinction between the corporeal and the physical domains resolves the seeming paradox; but whereas the distinction between a corporeal instrument I and the associated physical instrument SI renders state vector collapse conceivable, it does not tell us how actually to conceive of it. This is the question, then, that remains to be addressed.

Let us consider what happens in the process of measurement: a particle, or set of particles, emanating from the object, enters the corporeal instrument, and becomes in effect a part of the instrument. It is on account of this incorporation that the instrument registers the outcome of the measurement. This outcome is consequently the result of a vertical act which may be conceived as an act of primary causation mediated by a form.[12] The problem now is to understand how this act affects the quantum-mechanical system comprised of the physical object O plus the physical instrument SI. For this is indeed where the puzzle resides: According to quantum theory, O+SI constitutes a physical system, which should evolve in accordance with the Schrödinger equation, as physical systems normally do; why then does this not happen? We have already given a partial answer: What distinguishes O+SI is the fact that SI happens to be subcorporeal; but what effect does this have on the state of the composite system? The effect is as follows: Certain particles originally belonging to O belong later to SI, which entails—by the 'principle of de-superposition'—a restriction of their admissible states. The composite system O+SI, and hence its state vector, experiences therefore a discontinuity at the moment of measurement, and this is none other than state vector collapse.

Meanwhile it has become apparent that this collapse does not abrogate the Schrödinger evolution of the system, but merely 're-initializes' the Schrödinger equation. In other words, the change in trajectory results, not from a breakdown of temporal causality, but from an instantaneous change in the system itself; as always, vertical

12. It has been suggested by Eugene Wigner and some other quantum reality theorists that state vector collapse involves the 'consciousness' of a human observer. This hypothesis can, in my view, be safely discarded: it would be to overestimate the prowess of the physicist to suppose that he is able to collapse a state vector by an act of his mind. What is more, once the ubiquity of vertical causation has been grasped, there is no further need for such an *ad hoc* hypothesis.

causation does not impede horizontal modes of causality. It is therefore misleading to speak of 'chance' with reference to the microworld; what state vector collapse betokens is not randomness, not the toss of a die, but simply the fact that the spatio-temporal universe does not constitute a closed system. What is remarkable about the phenomenon is that it exhibits an effect of vertical causation, in defiance of the prevailing naturalism.

❖   ❖   ❖

It appears that a single 'principle of de-superposition' suffices to resolve the major enigmas of quantum physics:[13] the fact that a corporeal object X 'acts upon' SX to restrict the range of allowable superpositions explains at one stroke the phenomenon of state vector collapse as well as the Schrödinger paradox. Yet it is not in reality a question of X acting upon SX, but rather of a vertical act by which X itself is held in existence. What ultimately stands at issue is nothing less than the ubiquity of the cosmogenetic Act: this is what I would like now to explain.

To this end let us recall that all traditional cosmogonies envisage a bringing forth of the cosmos out of a primordial material substrate alluded to through a variety of symbolic forms in the sacred literatures of mankind, and later designated by various technical terms, from the Vedantic *Prakriti* to the Scholastic *materia prima*. Among all these designations of the material substrate, the one that is in a way most directly pertinent to our present inquiry is the Greek term *Chaos*; as we read in Hesiod's *Theogony*: 'Verily in the beginning, Chaos came to be.' What first 'came to be' may thus be conceived as a plethora of warring possibilities: 'warring' on account of being mutually incompatible on the plane of manifestation. A block of marble contains innumerable forms potentially; yet only one of these forms can be actualized by the sculptor's art. The

---

13. With the exception of 'nonlocality', an exceedingly enigmatic phenomenon, touched upon on pp 78–81. I have dealt with this matter from a traditional point of vantage in an article entitled 'Bell's Theorem and the Perennial Ontology'; see *Ancient Wisdom and Modern Misconceptions*, op. cit., chap. 3.

actualization of a form, it appears, requires a determinative act, the imposition of a Bound upon the Unbounded in accordance with the Biblical verse: 'He set His compass upon the face of the deep.' (Prov. 8:27) The cosmogenetic Act can therefore be conceived as an act of mensuration in the ancient sense common to both the Greek and Hindu traditions; as Ananda Coomaraswamy explains:

> The Platonic and Neoplatonic concept of measure agrees with the Indian concept: the 'non-measured' is that which has not yet been defined; the 'measured' is the defined or finite content of the universe, that is, of the 'ordered' universe; the 'non-measurable' is the Infinite, which is the source of both the Indefinite and of the finite, and remains unaffected by the definition of whatever is definable.[14]

In light of these traditional conceptions one sees once again that the quantum world occupies a position intermediate between the 'measured' and the 'non-measured': for whereas a quantum system is evidently subject to certain determinations—failing which it could not be conceived quantum-mechanically—it is yet insufficiently determined to qualify as 'a definite or finite content of the universe.' As we have noted before, it is not in reality a 'thing', which is to say that it lacks quiddity, lacks essence.

Now, it is precisely this lack of essence that manifests itself physically as quantum indeterminacy: herein, I say, lies the metaphysical significance of that indeterminacy. What has so greatly puzzled the physicist is simply a mark of the 'non-measured'. That mark, however, proves to be characteristic of the entire quantum world: the Heisenberg Uncertainty Principle guarantees as much. It follows that the quantum domain in its entirety constitutes a material substrate in relation to 'the measured', that is to say, the corporeal world. To be sure, a quantum system can indeed be actualized through what we have termed presentation or through measurement in the scientific sense: but it is to be noted that actualization inevitably takes us *out* of the quantum world and into the corporeal, while the

14. Quoted in René Guénon, *The Reign of Quantity* (Hillsdale, NY: Sophia Perennis, 2004), p 27.

system itself remains unmanifested and indeed unmanifestable. Do what one may, the substrate never ceases to be a substrate.

It appears from these considerations that quantum physics has discovered an ontological level approaching the primordial 'waters', which remain in place even after the Spirit of God has 'breathed upon them' to bring forth our world. I contend that quantum indeterminacy—the partial chaos of quantum superposition—can indeed be viewed as reflective of the primordial *Chaos*, or even more concretely as a remnant of this underlying 'disorder'.

As regards the actualization of a quantum system through measurement, we have seen that this hinges upon an act of vertical causation which is ultimately to be referred to primary causality. It can consequently be said that every measurement of a quantum system constitutes a cosmogenetic act which 'participates' in the single Act of creation. Whether the physicist realizes it or not, in the phenomenon of state vector collapse he is 'picking up' the cosmogenetic Act, not hypothetically, in some stipulated explosion that is supposed to have occurred so many billion years ago, but actually, in the *here* and *now*.

❖　❖　❖

One sees that SX represents the material substrate of X, its ontological underside, so speak. It constitutes, if you will, the black circle within the white field, the residual potency that refuses to be exorcised. This brings us back to a point I have made in Chapter 5: that indetermination represents 'the *yin*-side of the coin'. I will mention in passing that this *yin*-side, 'inexistent' though it be, plays a crucial role in the workings of the universe, from the behavior of inanimate objects to that of living organisms, and even, it seems, of civilizations. It is due to this 'black field' that growth, spontaneity, and a certain freedom exist in the world; yet, at the same time, the material substrate manifests itself also as a universal propensity towards decay, a tendency, if you will, to return to the primordial chaos. In the moral and social spheres this proclivity is of course familiar to us all,[15] while, from a scientific point of view, that same universal

15. The previously quoted book by Robert Bolton, entitled *The Order of the*

tendency manifests itself most directly in the Second Law of thermodynamics.[16]

While the existence of a material substrate—or 'sub-existential chaos'—was never in doubt, I find it truly astounding that a way has been found to represent that substrate physically and deal with it in a mathematically precise and empirically effective manner. It is safe to say that no one could have foreseen such a scientific development; and as a matter of fact, even after the event there appear to be few today who grasp its true significance.

It turns out that the physicist's search in quest of 'matter'—his centuries-long travail to ascertain the material basis of corporeal existence—has at last been crowned with success; it is only that he has so far failed to recognize the fact. Misled by his Cartesian presuppositions, he has been searching for *res extensae*, for Democritean atoms, if you will; and when, in the early decades of the twentieth century, it appeared that success was finally at hand, he found, in the decisive moment of discovery, that his quarry had mysteriously eluded his grasp. In place of *res extensae* the elusive quantum particles have made their appearance, and he was forced to concede, to his consternation, that these 'particles in name' are not in fact real entities, not truly 'things'. Yet the fact remains that the scientific *modus operandi* of his inquiry was sound, and that the long and arduous path of discovery has actually led to the material substrate of corporeal things, despite spurious philosophical appearances to the contrary. From a traditional point of vantage it is evident that SX does represent the substrate of X, and that in fact a more marvelous depiction of that 'underlying chaos' can hardly be conceived; but unfortunately that breakthrough has apparently been lost on the greater physics community, which still fondly imagines that modern science has disqualified the wisdom of the ages.

There is a moral in this story, with which it may be fitting to close

---

Ages, provides what may be the best introduction to this subject. See especially chaps. 9 and 10.

16. It is evident that Entropy, which is indeed a form of 'disorder', can hardly be unrelated to the primordial 'disorder' known to tradition as *Chaos*.

this monograph. Hard science, as I have noted elsewhere,[17] is ultimately self-corrective, and wiser, in a way, than the scientists who pursue it; in the end it is apt to lead us to the truth, if only we have 'eyes to see'. But science itself cannot give us this vision: science as such cannot interpret its own findings; and neither, I would add, can modern philosophy. What is called for, I maintain, is a grounding in the traditional metaphysical doctrines of mankind: the very tenets that have been decried since the Enlightenment as primitive, pre-scientific, and puerile. Strange as it may seem to modern minds, these teachings—like the vertical causality we have contemplated in this chapter—derive ultimately 'from above': from the Center of the circle, if you will. Originally formulated in the language of myth, they have served as a catalyst of metaphysical vision down through the ages; neither Plato, nor Aristotle, nor Aquinas invented their own doctrines: all have drunk from this spring—except, of course, for the pundits of modernity, who have rejected that heritage. By now, to be sure, one knows very well to what destination modernity leads: we have, after all, entered the disillusioned and skeptical era of postmodernism. The argument against the traditional wisdom has now run its course, and the way to the perennial springs is open once more. The time is ripe for a new interpretation of scientific findings based upon pre-Cartesian principles; what is called for is a radical change of outlook, a veritable *metanoia*. Whether the doctrines of science will conduce to human enlightenment or to the blighting of our intellect hangs in the balance.

17. 'Science and the Restoration of Culture,' *Modern Age*, vol. 43, no. 1 (2001).

APPENDIX
QUANTUM THEORY:
A BRIEF INTRODUCTION

THERE IS HARDLY A BETTER WAY to broach the subject of quantum theory than by reflecting upon the results of the so-called double-slit experiment, the idea of which is very simple indeed. Light, or some other kind of radiation, is made to pass through two slits in a screen S to a second screen R, where the intensity of the incoming radiation is somehow recorded or observed. For example, one may use sunlight, and observe the resultant intensity distribution at R by visual means; and as a matter of historical fact, it was in this form that the double-slit experiment was first performed in 1803 by an English scientist named Thomas Young. Now, as might be expected, when one slit is open and the other closed, one obtains a single bright line at R, positioned behind the open slit.[1]

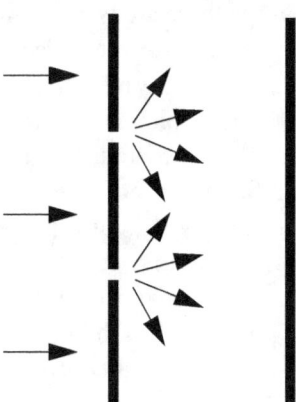

1. This will be the case provided the width of the slit is large compared to the wave length of light, or better said, the wave lengths falling within the visible range.

When both slits are open, on the other hand, one does not obtain simply two bright lines (one corresponding to each of the two open slits), but rather a pattern of bright and dark bands, the intensity of which diminishes gradually as one moves away from the slits in either direction. This is the first experimental result; and it can be readily explained in terms of a wave model for light. Consider a position P on the screen R, and let A and B denote the positions of the two slits.[2] Clearly, if the wave trains from A and B arriving at P are 'in phase', they will reinforce one another, and if 'out of phase', they will diminish or cancel. Now, whether the two wave trains are in phase or out of phase at P depends upon the distances AP and BP; and it is not hard to see, moreover, that 'in phase' and 'out of phase' positions will alternate as P moves across the screen R in a direction perpendicular to the slits. This results in a sequence of bright and dark bands, the very pattern obtained in the double-slit experiment.

It is to be noted that this effect has nothing especially to do with light, but applies to all kinds of wave propagation. The double-slit experiment, thus, could equally well be carried out with water waves, for example, which once again will reinforce each other when crest meets crest, and cancel when crest meets trough. In place of bright bands at R one now encounters positions at which the water rises and falls considerably, and in place of dark bands one has positions of relative calm. Wherever there are waves of any kind, there will be these so-called diffraction or interference phenomena, resulting from the fact that intersecting wave trains reinforce or cancel each other, depending upon their phase.

There is, however, a second experimental finding to be considered. Suppose one were to refine the procedure of Young's experiment so as to render the detection of light at R more accurate than is normally possible by visual means; one would then find that light is received, not continuously (as demanded by the wave model), but in discrete units. Thus, what one 'sees' by means of sufficiently sensitive instruments are tiny flashes of light, distributed at random

2. For the sake of simplicity we shall assume that the width of the slits, though large in comparison to the wave length of light, is yet small enough to be neglected in the calculation of interference effects.

over the screen R, but concentrated in the regions which, to the human eye, appear as bright bands. The picture that emerges, thus, when we refine our instruments of detection, is far more suggestive of a stream of particles, of tiny 'bullets of light', than it is of a continuously distributed wave.

Meanwhile it has been found that the so-called particles out of which atoms and molecules are made up (such as electrons and protons) likewise give rise to interference effects. In fact, it is possible to run the double-slit experiment with any of these so-called particles, and in each case one finds that the density of particle impacts at R exhibits the familiar diffraction pattern of a wave. It appears that the entities we are dealing with behave in certain respects like particles, and in other respects like waves.

What, then, is an electron, for example: is it really a particle, or is it actually a wave? Obviously it cannot be both; for it is manifestly impossible for a thing to be localized within a minute volume and at the same time be spread over a vast region of space. Now, the idea that an electron is simply a wave can be ruled out immediately. For if it were indeed a wave, it would evidently affect the screen R, not just at a particular point P (or immediate vicinity thereof), but at all points where the amplitudes of merging wave trains emanating from A and B do not cancel. But as we have seen, the electron does impact the screen R at a particular location, at which point it gives up its entire momentum (or kinetic energy) all at once.

An electron, therefore, is definitely not a wave. But even though it gives rise to diffraction effects, could it not still be a particle? Let us consider the matter. For the sake of simplicity (and to forestall the conceivable supposition that the diffraction phenomena might be a 'crowd effect') we will assume that the electron beam has been rarefied to the point where only one electron passes the screen S at a time (a condition which can indeed be achieved experimentally). Now, if slit A is open and B closed, each electron that passes through A will impact the screen R within some narrow band behind A; and the case will be similar if B is open and A closed.[3] If both slits

3. We must assume that the width of the slits is large compared to the so-called de Broglie wave length of the electron, i.e., the ratio h/p, where h is Planck's constant

are open, on the other hand, we find, once again, that interference bands appear. Meanwhile it is immaterial whether these electrons are passed through S in a dense stream (say 10^{24} electrons per second) or one at a time (say one per hour), as we have supposed.

Obviously something very strange is going on. For if an electron is indeed a particle, it must pass either through A or through B; and if it passes through A, it ought to impact within the given band behind A, and if through B, then within the band behind B. The distribution of electron impacts when both slits are open ought therefore to be the sum of the respective distributions when either slit is closed, which is to say that it should consist of the two aforementioned bands. This is what would happen, obviously, if we were firing, not electrons, but ordinary bullets. But in fact it does not![4] In place of the expected pair of bands we find an indefinite sequence of lines, spreading out in either direction from the slits. It appears, thus, that even though the electron, presumably, passes either through A or B (but not through both!), its behavior is affected by the state of the other slit: whether it be open or closed. But how does the electron 'know' whether the other slit is open or not? By what means does it probe the surrounding space?

It has become evident that an electron cannot be simply an ordinary or 'classical' particle. At best it could be such a particle 'plus' something else, something that is *not* localized. Could it be, then, a particle plus a wave? The idea has in fact been proposed, and pursued with some success; and yet it turns out that the hypothesis of a 'pilot wave' has contributed little, if anything, to our understanding. Moreover, by virtue of the weird properties which this so-called wave would have to have, the resultant picture is in any case far from 'classical'.

When all is said and done, we find that 'quantum strangeness' cannot be explained—or explained away—by any classical *tour de*

---

and p the momentum of the electron. Otherwise diffraction effects from a single slit would come into play.

4. According to quantum theory, corresponding diffraction effects arise even in the case of 'large' objects, such as baseballs and bullets. It is only that for these objects the effects in question are scarcely observable, due to the fact that the de Broglie wave length is then exceedingly small.

force. The simple results, even, of our humble double-slit experiment, already defy explanations of this kind. What emerges is the realization that the so-called fundamental particles of Nature are in fact neither particles nor waves in the strict sense, nor indeed are they entities of a more complex kind that could yet be conceived in the terms of classical physics. What is needed is a reformulation of the very foundations of physics: a new formalism, which radically breaks with the old. To be precise, it must be a formalism that distinguishes categorically between the quantum entity or system as such (an electron, for example) and its various observables (such as position, momentum, and so forth). The problem with classical physics, one can say, is that it conceives of physical entities in terms of their observables, thus rendering 'concrete' what in Nature is not yet so. It assumes, for example, that an electron must have a well-defined position and a well-defined momentum at all times, whether that position and momentum have been ascertained by measurement or not. But how do we know that? And is it necessary, moreover, to assume that Nature, in all of her operations, is thus constrained?

Now, the remarkable thing, perhaps, is not so much the fact that 'concrete' or 'classical' descriptions of physical reality must be given up, but that they *can* be—without putting an end to physical inquiry. The amazing thing, in other words, is that it is possible to transact the business of physics in terms of a formalism which distinguishes categorically, as I have said, between the physical system as such and its observables: a formalism, one might say, based upon an 'abstract' conception of physical reality.

The decisive step, let it be noted, was taken in 1925 by Werner Heisenberg, then 24 years old, when he hit upon a brand new way of representing physical systems. In what can only be termed a singular stroke of genius, the young Heisenberg conceived of the idea of representing a quantum system by an element or vector in what mathematicians are wont to term a Hilbert space: a mathematical structure which had evolved in the study of certain kinds of equations, and which by that time was quite well understood. One might add that Heisenberg, at the time of his decisive discovery, was entirely unacquainted with this mathematical development; in essence, thus, he re-invented the idea of a Hilbert space. Within a short period, however, the existing mathematical theory was pressed into the service of quantum mechanics; and thus the new physics obtained a solid and coherent mathematical foundation.

What I propose to do in this 'brief introduction' is, first of all, to explain to the general reader what kind of a thing a Hilbert space is, and then indicate how Hilbert spaces are employed in quantum theory. To keep the exposition as non-technical as possible, I will restrict myself to finite-dimensional spaces. My concern is to convey the basic facts in the simplest way possible, while relegating to the background all that might obscure the main ideas of this introduction, this first glimpse.

1. Finite-Dimensional Hilbert Spaces

We shall take as our starting point the familiar Euclidean plane: the plane, namely, in which the notions of distance and angle are defined. We will choose a point O in this plane; and having done so, we shall refer to points in the plane as vectors. What, then, is the difference between a vector and simply a point? It appears, at first glance, that the two are exactly the same! The difference, however, is this: Having chosen a reference point or so-called origin O, one is able to define three algebraic operations which depend upon this choice. The first is termed vector addition: two points (now called 'vectors') can be added so as to produce a third vector. The next operation is termed scalar multiplication: a vector can be 'multiplied' by an ordinary or so-called real number[5] so as to produce a second vector. And the last operation is called an 'inner product': two vectors can be thus 'multiplied' so as to produce a (real) number. The given set of vectors, endowed with the first two operations,

---

5. A real number is one that can be expressed in the familiar decimal notation. It is thus an integer (positive, negative, or zero), plus a number given by an expression of the form $.x_1x_2x_3 \ldots$, where the x_i's are 'digits' from the set $0, 1, 2, \ldots, 9$. Such an expression represents in fact an infinite series which converges to a real number between 0 and 1. In addition to the integers and integer fractions, the real numbers include the so-called irrational numbers, such as $\sqrt{2}$ and π.

constitutes an example of what is termed a vector space. Endowed with all three, on the other hand, it is more than just a vector space. It is now a (very small!) Hilbert space.

Let us indicate, first of all, how vector addition is defined (given a reference point O!). If P is a point in the plane, it will now be convenient to designate the corresponding vector by the notation $\overrightarrow{OP}$, which will no doubt be familiar to many readers. How, then, is one to define the sum of two vectors, $\overrightarrow{OP}$ and $\overrightarrow{OQ}$, let us say? We are given three points (O, P, and Q, namely), and must now look for a fourth point R which is somehow determined by the given three. Now, a natural choice would be the point R for which OPRQ constitutes a parallelogram; for indeed, by virtue of the geometric structure of the plane (the notion of parallelism, in this instance) the point R is uniquely determined by O, P, and Q.[6] Having determined R, moreover, an addition of vectors is then given by the formula

$$\overrightarrow{OP} + \overrightarrow{OQ} = \overrightarrow{OR}$$

And this is what one does: one defines vector addition by the so-called parallelogram rule.

Scalar multiplication is next. How is this to be defined? Here again a geometric notion comes into play, the concept of distance, this time. Given two points P and Q, we will denote the distance between them by the notation |PQ|. Now let $\overrightarrow{OP}$ be an arbitrary vector and α a real number. We will suppose, first of all, that α is positive. We would like scalar multiplication of $\overrightarrow{OP}$ by α to multiply the distance |OP| by α while keeping the direction of the line OP unaltered. To this end we observe that there exists a unique point R on the line OP such that

$$\alpha|OP| = |OR|,$$

and such that O, moreover, does not lie between R and P. For positive

---

6. We are assuming that the points O, P, and Q are not collinear. The reader may wish to ascertain for himself what is to be done in the collinear case.

values of α we are consequently at liberty to define the scalar product by the formula

$$\alpha \vec{OP} = \vec{OR}.$$

When α is negative, on the other hand, we can first multiply $\vec{OP}$ by $-\alpha$ (an operation which has already been defined), and then reverse the direction of the resultant vector $\vec{OR}$ (which now puts O between the points P and R). And finally, when α is zero, we shall take R to be O. The scalar product has now been defined.

The definition of the third algebraic operation (the inner product) hinges on the geometric notion of angle and may strike the reader as a bit more contrived. Suffice it to say that this product will be designated by the notation $\langle \vec{OP}, \vec{OQ} \rangle$ and defined by the formula

$$\langle \vec{OP}, \vec{OQ} \rangle = |OP| \, |OQ| \cos \phi$$

where ϕ denotes the angle between the lines OP and OQ. The reader will note that the right side of this formula represents a real number determined by the vectors $\vec{OP}$ and $\vec{OQ}$, as indeed it should.

All three algebraic operations have now been specified, and as might be expected, it turns out that they satisfy a number of rather simple algebraic rules (the very rules, in fact, in terms of which the structure of a Hilbert space is axiomatically defined).[7] For example, vector addition is commutative, which is to say that the order in which two vectors are added is immaterial. This algebraic property, incidentally, is obvious from the parallelogram rule. Less obvious, on the other hand, is the fact that vector addition is associative: that if we add three vectors, it matters not which two are added first. Another nice (and unobvious) property is this: Scalar multiplication is distributive with respect to vector addition, which is to say that the following formula holds in all cases:

---

7. What we have done is to translate geometric into algebraic properties. The algebraic structure of our Hilbert space 'mirrors' the Euclidean structure of the plane.

$$\alpha \, (\overrightarrow{OP} + \overrightarrow{OQ}) = \alpha \overrightarrow{OP} + \alpha \overrightarrow{OQ}$$

So as not to leave the inner product altogether out of account, let us mention, finally, that the latter is bilinear. This means that

$$\langle \alpha \overrightarrow{OP}, \overrightarrow{OQ} \rangle = \alpha \langle \overrightarrow{OP}, \overrightarrow{OQ} \rangle$$
$$\langle \overrightarrow{OP} + \overrightarrow{OQ}, \overrightarrow{OR} \rangle = \langle \overrightarrow{OP}, \overrightarrow{OR} \rangle + \langle \overrightarrow{OQ}, \overrightarrow{OR} \rangle$$

plus two similar properties with right and left interchanged.

Two observations need now to be made. Firstly, one can calculate (or if you will, define) the 'length' of a vector in terms of the inner product by the formula

$$|\overrightarrow{OP}| = \sqrt{\langle \overrightarrow{OP}, \overrightarrow{OP} \rangle}$$

And secondly, it makes sense to say that two vectors are perpendicular (or 'orthogonal', as mathematicians prefer to put it) if their inner product is zero. The reader will note that two nonzero vectors $\overrightarrow{OP}$ and $\overrightarrow{OQ}$ are orthogonal (in this sense) if and only if the lines OP and OQ are perpendicular.

We are now in a position to exhibit a formula which plays a major role in quantum theory. Let $\overrightarrow{OX_1}$ and $\overrightarrow{OX_2}$ denote mutually orthogonal vectors of unit length, and let $\overrightarrow{OP}$ denote an arbitrary vector. Applying the algebraic laws to which I have alluded, and making use of an elementary property of the plane, it is not hard to show that

$$(1) \quad \overrightarrow{OP} = \alpha_1 \overrightarrow{OX_1} + \alpha_2 \overrightarrow{OX_2}$$

where $\alpha_1 = \langle \overrightarrow{OP}, \overrightarrow{OX_1} \rangle$ and $\alpha_2 = \langle \overrightarrow{OP}, \overrightarrow{OX_2} \rangle$. The geometric meaning of this formula, moreover, becomes evident once it is recognized that the two terms on the right side of equation (1) correspond to the perpendicular projections of the point P onto the lines OX_1 and OX_2, respectively. Or to put it another way: they represent two sides of a parallelogram (a rectangle, in fact), of which OP is a diagonal.

The two mutually orthogonal unit vectors $\overrightarrow{OX_1}$ and $\overrightarrow{OX_2}$ are said to constitute an orthonormal basis for our Hilbert space. It is now to be observed that if we had started, not with the Euclidean plane, but with the full three-dimensional Euclidean space, we could have defined each of the three algebraic operations exactly as before, and these operations will in fact satisfy exactly the same algebraic rules. However, formula (1) will no longer hold. In place of two mutually orthogonal unit vectors, one now needs three (which again are called an orthonormal basis). One then obtains

$$(2) \quad \vec{OP} = \alpha_1 \vec{OX} + \alpha_2 \vec{OX_2} + \alpha_3 \vec{OX_3}$$

where for $\alpha_i = \langle \vec{OP}, \vec{OX_i} \rangle$ for $i = 1, 2, 3$.

One sees from the two examples that an orthonormal basis is characterized not only by the condition that the given vectors are mutually orthogonal and of unit length, but also by the fact that the set is 'maximal' in the sense that it is impossible to add another unit vector to the set which will be orthogonal to each of the given vectors. With this understanding it can then be shown that any two orthonormal bases of a given Hilbert space must contain the same number of vectors; and that number defines the *dimension* of the Hilbert space.

What I have exhibited thus far are Hilbert spaces of dimensions 2 and 3, respectively. I need to point out next that it is easily possible to construct Hilbert spaces of dimension n for every positive integer n. It is true, of course, that for $n > 3$ one is no longer able to visualize these vector spaces in a concrete geometric way; but yet all the familiar rules apply, and it is in reality as easy to work in these higher-dimensional spaces as in our 2 and 3-dimensional examples. It needs hardly to be pointed out that in dimension n equations (1) and (2) assume the form

$$(3) \quad \vec{OP} = \alpha_1 \vec{OX_1} + \alpha_2 \vec{OX_2} + \ldots + \alpha_n \vec{OX_n}$$

where $\alpha_i = \langle \vec{OP}, \vec{OX_i} \rangle$ for $i = 1, 2, \ldots, n$.

However, no matter how large we take the integer n to be, it turns out that these n-dimensional Hilbert spaces are still too small for most applications in quantum theory. What is needed, thus, are 'infinite-dimensional' Hilbert spaces; and admittedly, it is not possible to describe these in terms which readers untrained in mathematics could be expected to understand. But neither is it necessary to do so; for it happens that the main ideas of quantum theory can be very well explained in a finite-dimensional setting. The fact that in infinite dimensions the story becomes a good deal more complicated does nothing to alter the basic picture, as exemplified in the finite-dimensional case. On the contrary! What the intricacies of Hilbert-space theory (for example, the justly celebrated spectral decomposition theorem for Hermitian operators) tell us is precisely that the elementary picture does carry over in essence to the infinite-dimensional case.

What we propose to do in this introduction is to present the mathematical structure of quantum theory in a finite-dimensional setting, and thus in a simplified form.

2. Complex Numbers

The Hilbert spaces thus far considered have been real Hilbert spaces, meaning thereby that the 'numbers' or so-called scalars involved in scalar multiplication and the inner product are real numbers. It happens, however, that quantum theory demands complex Hilbert spaces: spaces in which the scalars are complex numbers. Formally, everything remains the same. One has, once again, the three algebraic operations, and these satisfy exactly the same rules as before. It is only that the underlying concept of number has been enlarged from the real to the complex field, as this kind of algebraic structure is called.

Paul Dirac (one of the founders of quantum theory) once remarked that 'God used beautiful mathematics in creating the world.' This would in any case explain the appearance of complex numbers in physics; for as every mathematician knows, mathematical analysis attains its perfection in the complex domain.

What, then, is a complex number? In certain textbooks one learns that it is a number of the form x + iy, where x and y are real numbers and i is said to be 'the square root of –1.' But then, how does one know that –1 *has* a square root, and that moreover this 'imaginary number' i can be multiplied by a real number y and the result added to another real number x? Clearly, the expression $x + iy$ is not a definition, but simply a notation. As such, however, it is useful; for the notation itself suggests that these 'numbers' should be added and multiplied according to the rules

$$(4) \quad (x + iy) + (x' + iy') = (x + x') + (y + y')$$

$$(5) \quad (x + iy)(x' + iy') = (xx' - yy') + i(xy + yx').$$

One can now verify quite easily that this addition and multiplication satisfy all the usual conditions, which is to say that these 'numbers' (if they exist!) constitute a field. Inasmuch as every real number x is also a complex number (one for which $y = 0$), one sees moreover that this field 'extends' the field of real numbers.

But the question remains: What is a complex number? Now, the simplest and most natural answer is this: A complex number is an ordered pair (x, y) of real numbers, with the proviso that addition and multiplication are defined by formulas analogous to (4) and (5), namely,

$$(x, y) + (x', y') = (x + x', y + y')$$

$$(x, y)(x', y') = (xx' - yy', xy' + yx').$$

Let me note, firstly, that real numbers can again be viewed as a special case of complex numbers [by 'identifying' x with the pair $(x, 0)$], and that the present definition of complex numbers at once resolves the enigma of i, the 'imaginary' square root of -1. For the given rule of multiplication tells us immediately that

$$(0, 1)(0, 1) = (-1, 0),$$

which shows that i is none other than the complex number $(0, 1)$. One sees moreover, that it is quite misleading to speak of 'real' and 'imaginary' numbers, for the pair $(0, 1)$ is obviously no more 'imaginary' or less 'real' than $(1, 0)$.

I mention in passing that complex numbers can be readily represented by vectors in a 2-dimensional space, which is to say that one can think of complex numbers as points in a plane (the so-called complex plane). For future reference, every complex number has a so-called absolute value (its distance from the origin in the complex plane), given by the formula

$$|(x,y)| = \sqrt{x^2 + y^2}.$$

Complex numbers of absolute value 1 constitute thus a circle in the complex plane. These numbers can consequently be coordinatized by an angle: the angle θ, let us say, subtended in a counter-clockwise sense from the half-line consisting of the positive real numbers. And I note (again for future reference) that the complex number on the unit circle corresponding to the angle θ (which we will take, not in degrees, but in so-called radians) is given by the notation $e^{i\theta}$. The fact that $e^{i\theta}$ is actually the real number e (the so-called base of the natural logarithms) raised to the 'imaginary' power $i\theta$ need not concern us.

3. State Vectors and Observables

With every physical system, quantum mechanically conceived, there is associated a complex Hilbert space, the nonzero vectors of which correspond to states of the physical system. These vectors are referred to as state vectors, and following Dirac, are generally denoted by a Greek letter, with a vertical bar on the left and a bracket on the right. The inner product of $|\psi\rangle$ and $|\chi\rangle$, let us say, will be written as $\langle \psi, \chi \rangle$.

Now let $|\psi\rangle$ and $|\chi\rangle$ be state vectors, and let α and β denote complex numbers. The weighted sum $\alpha|\psi\rangle + \beta|\chi\rangle$ is then another vector in the Hilbert space. But recall that nonzero vectors in our Hilbert space correspond to states of the physical system! The complex weighted sum $\alpha|\psi\rangle + \beta|\chi\rangle$ has thus a physical significance (provided it is not zero): it represents a possible state of the system. This remarkable fact, which has no analogue in classical physics, is known as the superposition principle.

It should be mentioned next that multiplication of a state vector by a nonzero complex number does not change the corresponding state, which is to say that physical states correspond, not to individual state vectors, but to what could be termed a complex line through the origin of the vector space.

Let us now consider an observable of the physical system: some quantity, namely, which can in principle be ascertained by an act of measurement. The outcome of a measurement, of course, depends upon the state of the system. One needs, however, to distinguish between two cases. There are, first of all, states for which the outcome is determined with certainty. These are called eigenstates of the given observable. In general, however, the value of the observable will not be determined with certainty; which is to say that when the system happens not to be in an eigenstate, a measurement can in principle yield any of a number of possible values. The possible values of an observable, moreover, are called eigenvalues; and finally, state vectors corresponding to an eigenstate are termed eigenvectors.

We come now to a crucial fact: eigenvectors corresponding to different eigenvalues are orthogonal. This implies, in particular, that if the observable can assume n distinct values, and if each of these has an eigenstate, then the Hilbert space must be at least n-dimensional. And by the same token, if the number of distinct eigenvalues is infinite, and each has an eigenstate, then the Hilbert space must be infinite-dimensional.

For the sake of simplicity let us henceforth assume that the Hilbert space is finite-dimensional, say a space of dimension n. It follows then, on the strength of a mathematical theorem, that every observable admits an orthonormal basis of eigenvectors. Let us now choose an observable, and let $|\psi_1\rangle, |\psi_2\rangle, \ldots, |\psi_n\rangle$ denote such an orthonormal basis. Every state vector $|\chi\rangle$ can then be represented as a complex weighted sum of the given eigenvectors. To be precise, equation (3) gives

$$|\chi\rangle = \alpha_1|\psi_1\rangle + \alpha_2|\psi_2\rangle + \ldots + \alpha_n|\psi_n\rangle$$

where $\alpha_i = \langle \chi, \psi_i \rangle$ for $i = 1, 2, \ldots, n$. The question arises now whether the coefficients $\alpha_1, \alpha_2, \ldots, \alpha_n$ (which describe the position of the state vector in relation to the given eigenvectors) carry physical information of some kind. Since a state vector $|\chi\rangle$, however, can be multiplied by a nonzero complex number without altering the corresponding state, one sees that the α_i's are defined only up to a nonzero multiple. To remedy this lack of determination, one can 'normalize' the state vector $|\chi\rangle$ by dividing it by its length. The resultant α_i's are then uniquely determined, except for complex multiples of absolute value 1, and will moreover satisfy the condition

(6) $|\alpha_1|^2 + |\alpha_2|^2 + \ldots + |\alpha_n|^2 = 1.$

What, then, is their physical significance? It is this: the square of the absolute value of α_i (which is now uniquely determined) is precisely the probability that a measurement of the given observable in the given state will yield the eigenvalue λ_i corresponding to the eigenvector.[8] The probability p_i of obtaining the value λ_i is consequently given by the formula

$$p_i = \frac{|\langle \chi, \psi_i \rangle|^2}{\langle \chi, \chi \rangle}.$$

It is to be noted, first of all, that by virtue of equation (6) these probabilities add up to 1, as indeed they should. Let us suppose, now, that $|\chi\rangle$ is an eigenvector of the observable. To be concrete, let us suppose it is $|\psi_1\rangle$. It follows now that $\alpha_1 = 1$, and all remaining ai's are zero. But this means that a measurement of the given observable will yield the eigenvalue λ_1 with probability 1, that is to say, with certainty. One thus recovers what has been said earlier concerning the outcome of measurements when the system is in an eigenstate.

In general, however, the system will be in a superposition of eigenstates, in which case the state determines, not the actual outcome of a measurement, but only the probabilities associated with possible outcomes. It has been debated since 1925 whether this 'indeterminacy'

---

8. We are assuming, for the sake of simplicity, that the eigenvalues $\lambda_1, \lambda_2, \ldots, \lambda_n$ are all distinct. For a multiple eigenvalue λ the probability turns out to be the sum of the probabilities p_i associated with the eigenvectors $|\psi_i\rangle$ corresponding to λ.

is due to the limitations of quantum theory, or whether indeed 'God plays dice', to put it in Einstein's famous words.

4. The Heisenberg Uncertainty Principle

A few words, at least, should now be said concerning the celebrated Heisenberg uncertainty principle. Given two observables P and Q, the question arises whether the values of both can be determined with certainty. In light of what has been said before, one sees that for this to be the case, the system would have to be in an eigenstate of both P and Q. In general, however, an eigenvector $|\chi\rangle$ of Q will be a weighted sum of eigenvectors $|\psi_1\rangle, \ldots, |\psi_n\rangle$ belonging to P, the coefficients of which are simply the inner products $\langle \chi, \psi_i \rangle$. And this means that if our system happens to be in an eigenstate corresponding to the eigenvector $|\chi\rangle$, a measurement of P can in principle yield any eigenvalue λ_i of P, provided only that the corresponding inner product $\langle \chi, \psi_i \rangle$ is not zero. The very fact that the value of Q is determined with certainty can imply, under these auspices, that the value of P is indeterminate.

In general, of course, the system will be in a state that is not an eigenstate of either P or Q, which is to say that the values of both observables will be indeterminate. There exists, however, a mathematical measure of indetermination (termed the standard deviation) in terms of which it is possible to state a relation between these two 'uncertainties.' This relation takes the form

$$(7) \quad \Delta P \Delta Q \geq \{P, Q\},$$

where ΔP and ΔQ denote the standard deviations of P and Q, respectively, and $\{P, Q\}$ denotes a certain non-negative number determined by P and Q. Formula (7) expresses the so-called generalized uncertainty relation. It affirms that no matter what state the system is in, the product of the two 'uncertainties' cannot be smaller than $\{P, Q\}$.[9] Now, the Heisenberg uncertainty principle, properly so

9. Roughly speaking, the more accurately P is known, the greater will be the uncertainty of Q.

called, corresponds to the special case where Q represents a position and P the corresponding momentum coordinate of a particle, or more generally, the case where Q and P are so-called conjugate observables, in which case (7) reduces to

$$\Delta P \Delta Q \geq h/2\pi,$$

where h is Planck's constant.

All this, however, is of little immediate interest to us. What matters, for the purpose of this cursory introduction, is the simple recognition that a quantum system can never be in a state for which the values of all observables are determined with certainty. And this fact is implied by the very structure of quantum theory, that is to say, by the basic principles enunciated in the preceding section.

5. The Schrödinger Equation

The state of a physical system is obviously subject to change. State vectors, therefore, must in general depend upon a time-coordinate t, and when necessary, we will indicate this functional dependence by the notation $|\psi(t)\rangle$. The question arises now whether it is possible to predict the future values $|\psi(t)\rangle$ of a state vector, given an initial value $|\psi(t_0)\rangle$. For this to be case, it is of course necessary to make appropriate assumptions regarding the action of external forces upon the given physical system. In technical parlance, one assumes that these forces are conservative, that is to say, derive from a so-called potential; and we shall henceforth assume that this condition is satisfied. Does there exist, then, an equation that enables us to calculate the future values of a given state vector?

The desired equation was discovered in 1926 by the Austrian physicist Erwin Schrödinger. It affirms, first of all, that state vectors evolve linearly. This means that if a relation

$$|\chi\rangle = \alpha_1 |\psi_1\rangle + \alpha_2 |\psi_2\rangle$$

(with given complex coefficients α_1 and α_2) between state vectors holds at some time t_0, it will continue to hold for all $t > t_0$.

Let us again suppose that we are in an n-dimensional Hilbert space, and let $|\psi_1\rangle, \ldots, |\psi_n\rangle$ be a set of state vectors which, at time $t = t_0$, constitute an orthonormal basis. At time $t = t_0$, an arbitrary state vector $|\chi\rangle$ can then be expressed in the form

$$|\chi(t_0)\rangle = \alpha_1|\psi_1(t_0)\rangle + \alpha_2|\psi_2(t_0)\rangle + \ldots + \alpha_n|\psi_n(t_0)\rangle.$$

By the linearity of Schrödinger evolution, however, this implies that

$$(8) \quad |\chi(t)\rangle = \alpha_1|\psi_1(t)\rangle + \alpha_2|\psi_2(t)\rangle + \ldots + \alpha_n|\psi_n(t)\rangle$$

for all $t \geq t_0$.

But this equation enables us to calculate the Schrödinger evolution of an arbitrary state vector $|\chi\rangle$, once the Schrödinger evolution of the given basis is known! And this leads, obviously, to the question: Can we find a special basis for which the Schrödinger evolution assumes a particularly simple form, a form which can be easily ascertained?

The crucial fact is that eigenstates of the total energy (which is always an observable of the system) turn out to be stationary states: states which do not change at all. The fact that energy eigenstates are stationary, however, does not imply that energy eigenvectors are constant; for if that were the case, it would follow by equation (8) that all other state vectors are constant as well! The point is that state vectors can be multiplied by a nonzero complex number without altering the corresponding state. The Schrödinger evolution of an energy eigenvector must consequently be given by a complex factor, a certain complex function of time. What, then, is that function? It turns out to be

$$(9) \quad e^{-2\pi iEt/h}$$

where E is the given energy eigenvalue and h is Planck's constant. This represents a unit vector in the complex plane, which moreover rotates in a clockwise sense, with a frequency E/h. Energy eigenvectors, thus, engage in a ceaseless rotatory motion, the frequency of which is proportional to the corresponding energy.

Now let us assume that our orthonormal basis consists in fact of energy eigenvectors. The Schrödinger evolution of that basis is then given by the equations

$$(10) \quad |\psi_j(t)\rangle = e^{-2\pi i E_j t/h} |\psi_j(t_0)\rangle$$

for $j = 1, 2, \ldots, n$, where E_j denotes the energy eigenvalue corresponding to the eigenvector $|\psi_j(t)\rangle$. Substituting these expressions into equation (8), one obtains a formula for the Schrödinger evolution of $|\chi\rangle$.

This equation exhibits $|\chi\rangle$ as a superposition of simple oscillations, much as an arbitrary tone can be represented as a superposition of pure tones. It is however to be noted that the 'vibration' or 'wave motion' described by equation (10) pertains necessarily to a sub-empirical level (if indeed it constitutes a real process at all); for the constituent state vectors $|\psi_j(t)\rangle$ belong to one and the same physical state, and therefore cannot be distinguished by empirical means. But though the factor (9) proves thus to be unobservable, it nonetheless determines the Schrödinger evolution of every state vector in the Hilbert space. All the interference effects of quantum theory, moreover, hinge upon this mysterious complex oscillation, this so-called phase factor. It controls everything, but itself eludes scrutiny.

6. Schrödinger Evolution and State Vector Collapse

One of the basic facts of classical physics is that an initial state of a physical system determines its future states, provided only that the external forces acting upon the system are known in advance.[10] A universe governed by the laws of classical physics would therefore be deterministic: the course of its entire evolution, down to the smallest

---

10. This holds true in the Hamiltonian representation, which conceives of a physical system as a point in what is termed phase space: a space coordinatized by the positions and momenta of all constituent particles. The phase space of a system consisting of n particles is consequently 6n-dimensional.

details, would be uniquely determined from the first moment of its existence. We should not, however, be unduly surprised to find that things are not quite as simple in the case of quantum theory.

There exists, first of all, the Schrödinger equation, which enables one to predict the future states of a quantum system from an initial state. And one might add that Schrödinger arrived at his equation by way of classical physics, firm in the conviction that the familiar determinism would carry over into the quantum domain. And it almost does! For the most part state vectors do sweep out a continuous trajectory in Hilbert space, in accordance with the demands of the Schrödinger equation.[11] It happens, however, that this continuous and predictable evolution is interrupted occasionally by certain special events, which may cause the state vector to change abruptly and unpredictably: to 'jump', as the expression goes. What, then, causes these sudden jumps? It is none other than the act of measurement, the actual experimental determination of some given observable. Apparently, it is the intervention of the experimental process that causes the physical system to jump, to change instantly from one state to another, without passing through a continuous array of intermediate states (in accordance with Schrödinger's equation).

Let a physical system and an observable of that system be given. For simplicity, we will consider only what are termed experiments of the first kind: experiments for which the observable in question assumes its measured value at the termination of the measurement.[12] Now, if a measurement yields the eigenvalue λ (by an experiment of the first kind), then we know that the observable has the value λ at the

11. Inasmuch as a single nonzero vector in Hilbert space determines thus the entire trajectory on which it lies, it follows that an arbitrary initial state determines the Schrödinger evolution of the quantum system. All this, of course, on condition that the external forces are determined in advance.

12. The point is that there are experiments ('of the second kind') which alter the value of the observable that is being measured. For example, one often determines the momentum of a nuclear particle by measuring the momentum transfer in a collision with some other particle. The momentum of the particle in question is consequently changed by the measurement. And thus, if a second measurement were to be made immediately after the first, it would give a different result.

termination of the measurement, which is to say that the system is at that moment in an eigenstate corresponding to the eigenvalue λ. Prior to the measurement, on the other hand, the system will in general have been in some superposition of eigenstates. It follows that the state vector has undergone a discontinuous change: a so-called collapse. By the act of measurement it has been cast, as it were, into an eigenstate of the given observable. One cannot say in advance, moreover, which eigenstate it will be, for as we have seen, quantum theory gives only probabilities in that regard. In general, therefore, the act of measurement gives rise to an unpredictable discontinuity which interrupts the deterministic Schrödinger evolution of the state vector.

No one seems to understand why the particular interactions which we term measurements have this remarkable effect. What exactly is it that differentiates a measurement from any other kind of interaction? Or to put it still more simply: Why do state vectors collapse? And most importantly, does that so-called collapse betoken an actual indeterminacy in the operations of Nature? As one knows, these questions have been pondered and vigorously debated ever since quantum theory saw the light of day; but up till now, at least, it appears that no definitive answers have been found. A majority of physicists, meanwhile, seem to be content to regard the duality of Schrödinger evolution and state vector collapse simply as a scientific fact of life; it is something which, by force of necessity, the working physicist learns to accept without too much question.

7. The Wavefunction of a Particle

Let us again assume that we are in an n-dimensional Hilbert space, and let $|\psi_1\rangle, |\psi_2\rangle, \ldots, |\psi_n\rangle$, denote an orthonormal basis of eigenvectors belonging to some observable of the physical system. Since an arbitrary state vector $|\chi\rangle$ can now be written as a weighted sum of the given eigenvectors, with coefficients $\alpha_i = \langle \chi, \psi_i \rangle$, one sees that $|\chi\rangle$ can be represented by the complex n-tuple $(\alpha_1, \alpha_2, \ldots, \alpha_n)$. For every observable there is consequently a corresponding way of representing state vectors by complex n-tuples. Let us now suppose that our system consists of a single particle, which is free to assume n positions, coordinatized by real numbers $x_1, x_2, \ldots, x_n$. The x_i's, then, are the eigenvalues of a particular observable. Now let $|\chi\rangle$ be a state vector and $(\alpha_1, \alpha_2, \ldots, \alpha_n)$ the n-tuple representation of $|\chi\rangle$ corresponding to this observable. One can now define a complex-valued function ψ on the set of position eigenvalues by the formula $\psi(x_i) = \alpha_i$, for $i = 1, 2, \ldots, n$. And this function ψ is called the wavefunction of the given particle.

One is especially interested, of course, in the case of a particle which is free to assume all positions in some three-dimensional region V of space. Now, such a particle evidently requires an infinite-dimensional Hilbert space for its quantum mechanical description, and it happens that an orthonormal basis of position eigenvectors does not exist in this case. Our previous construction of a wavefunction, therefore, no longer applies. It is possible, however, to proceed by other means, and one finds that state vectors can again be represented by a wavefunction, which is now a continuous complex-valued function ψ on V.

In general, a wavefunction is said to be normalized if the corresponding state vector is normalized, that is to say, has unit length. The reader will observe that for a normalized wavefunction ψ in the finite-dimensional case, $|\psi(x_i)|^2$ is precisely the probability of finding the given particle at x_i. The corresponding quantity $|\psi(x)|^2$ in the infinite-dimensional case, on the other hand, is not, strictly speaking, a probability, but a so-called probability density. It tells us that the probability of finding the particle within a "small" volume ΔV around x is given by $|\psi(x)|^2 \Delta V$.

One further remark about wavefunctions: it is not hard to see that the wavefunction corresponding to a weighted sum of state vectors is none other than the corresponding weighted sum of wavefunctions. And this implies that a weighted sum of wavefunctions is again a wavefunction.[13]

13. Wavefunctions, therefore, satisfy their own superposition principle; in fact, they constitute a Hilbert space. I will point out, in this connection, that quantum mechanics was discovered twice: first by Heisenberg, who based his theory on the Hilbert space of state vectors, and a short time later (and independently), by

8. *The Double-Slit Experiment Reconsidered*

Let us now return to the remarkable experiment which we have considered at some length in the beginning. A particle (an electron, let us say) is fired through a screen S endowed with two slits, and impacts a second screen R. Our physical system consists now of a single electron, subject to the prescribed conditions. If slit A is open and B closed, the electron is known to pass through A. Its wavefunction ψ_A, at that moment, will consequently be concentrated or 'peaked' at the slit A, which is to say that the amplitudes $\psi_A(x)$ will be zero for positions x away from the slit. Similarly, if B is open and A closed, the corresponding wavefunction ψ_B will be peaked at B at the instant the electron passes through that slit.

Let us now form a weighted sum

$$\psi = \alpha\psi_A + \beta\psi_B$$

of these two wavefunctions, with nonzero complex coefficients α and β. By what has been said above, ψ is again a wavefunction. And that wavefunction is descriptive of the case where both slits are open. The electron is now in a 'state of superposition', the kind of state that exhibits the previously considered interference effects (which, as we have seen, prove to be inexplicable in classical terms).

We may assume, without loss of generality, that the wavefunctions ψ_A, ψ_B, and ψ are all normalized, so that the squared absolute

Schrödinger, who based his theory on the Hilbert space of wavefunctions. It was Schrödinger, moreover, who demonstrated the equivalence of the two theories by establishing a so-called isomorphism between the respective Hilbert spaces (which in fact reduces, in the finite-dimensional case, to the correspondence between state vectors and wavefunctions, as given above). However, by virtue of the fact that the Schrödinger formalism gives preference to a particular observable (i.e., position in space), it is far less abstract than the formalism of Heisenberg, so much so that it lends itself more readily to a classical interpretation, which turns out, however, to be untenable. Schrödinger himself, oddly enough, looked upon the wavefunction from a classical point of view. And when Bohr explained to him, one day, the inevitability of wavefunction collapse, he gave the famous reply: 'If I had known of this damned "jumping" I would never have involved myself in this business in the first place.' Like Einstein, Schrödinger never fully came to terms with quantum theory.

values of their amplitudes are actually probability densities. The fact that ψ_A is initially peaked at A tells us, then, that the electron passes through A; and the case is obviously similar for ψ_B. The probabilistic significance of ψ, moreover, is likewise clear: the fact that it is doubly peaked implies that there is a positive probability that the electron passes through slit A, and a positive probability that it passes through slit B.

Let us suppose, now, that the wavefunction ψ is known at the initial moment, the instant at which the electron passes through S. By means of the Schrödinger equation one can then calculate $\psi(t)$ for all subsequent values of the time-coordinate, right up to the moment when the electron impacts the screen R. And as might be expected, the resultant probability density of particle impacts at R does indeed exhibit the familiar interference bands. On the level of state vectors, after all, we are in fact dealing with a superposition of waves; which is to say that from a mathematical point of view, the given bands do indeed constitute a diffraction phenomenon in the classical sense.

The fact is that quantum theory explains the experimental findings perfectly. And it does so, as we have just seen, by way of complex amplitudes, replete with their oscillatory phase factors. What is actually observable, on the other hand, are the squared absolute values: the probabilities and probability densities, namely, as manifested, for example, by a density of dots on an exposed photographic plate. The question arises, therefore, whether the complex amplitudes as such betoken a physical reality. Some physicists doubt that this is the case. But then, under these auspices, one would be hard-pressed to understand how a calculation, based upon fictitious amplitudes, could invariably give correct results. To speak concretely: If the initial twin peaks of the superposition wavefunction ψ are not in some sense real, how, then, are we to explain the appearance of interference effects? If it be true that an effect must have a cause, then we are justified in regarding a wavefunction as something more than a fiction. And then, by the same token, we are obliged to conclude that an electron, prior to the moment when it is actually observed, is somehow spread out in space. And if it happens, moreover, that its wavefunction is initially doubly peaked, one is likewise forced to conclude that in a sense it does pass through both slits, strange as this might seem.

Meanwhile quantum theory as such has nothing to say regarding the ontological status of its complex amplitudes; it simply informs us how to calculate quantum mechanical probabilities, and for the rest, leaves us to think as we will.

GLOSSARY

Associated physical object (p34): Every corporeal (i.e., perceptible) object can be subjected to measurements and conceived in physical terms. The corporeal object X determines thus a physical object SX, which is termed the associated physical object.

Bifurcation (p8): The Cartesian tenet which affirms that the perceptual object is private or merely subjective. The idea of bifurcation goes hand-in-hand with the assumption that the external world is characterized exclusively by quantities and mathematical structure. According to this view all qualities (such as color) exist only in the mind of the percipient.

Corporeal object (p26): A corporeal object is simply a thing that can be perceived.

Corporeal world (p26): The familiar (or 'pre-scientific') world which we know directly by way of sense perception.

Display (p36): A mode of physical observation which terminates, not in a numerical value or quantity, but in some kind of graphic presentation.

Eigenstate (p62): A state of a physical system in which the value of a given observable can be predicted with certainty.

Eigenvector (p62): In the formalism of quantum theory, the state of a physical system is represented by a so-called state vector. A state vector is called an eigenvector (with respect to some given observable X) if the value of X can be predicted with certainty whenever the physical system is in a state corresponding to the given state vector.

Essence (p85): Essence is what answers to the question 'What?'; it is thus the 'whatness' or quiddity of the thing.

Forma (p85): The Scholastic equivalent of morphe in the Aristotelian sense. The forma or 'form' is what renders a thing intelligible.

Hyle (p84): A term used by Aristotle to refer to the pre-existential recipient of form or intelligibility. The Greek word means 'wood,' and the metaphor is sculptural: just as a piece of wood can receive the form of Apollo or Socrates, so hyle can receive morphe or 'form' in a general sense.

Materia (p85): The Scholastic equivalent of the Aristotelian hyle. Materia is thus the pre-existential recipient of 'form' in the sense of an intelligible content.

Materia secunda (p88): A partially determined recipient of form or determination.

Materia quantitate signata (p89): This term is used to refer to a recipient of form or determination that is itself subject to a mathematical form or structure.

Morphe (p84): The formal or intelligible aspect of an existent entity. The term was employed by Aristotle in conjunction with the word hyle (the recipient of morphe).

Natura naturata (p105): A Scholastic term signifying 'nature' in the sense of something which has been produced, created or 'natured'.

Natura naturans (p105): Nature conceived as an active, creative or 'naturing' principle. The term is in truth a nomen Dei, 'a name of God'.

Nature (p77): I employ this term provisionally, inspired by Heisenberg's observation that modern physics deals, not with Nature as such, but with 'our relations to Nature'. The notion is subsequently rendered more precise with the help of Aristotelian and Scholastic conceptions.

Physical object (p29): A thing that can be known through the modus operandi of physics.

Physical system (p32): A physical object as conceived in terms of a given representation.

Physical universe (p29): The locus or domain of physical objects, and thus, in a way, the world as conceived by the physicist.

Potentia (p62): An Aristotelian term signifying something that exists 'in potency' in relation to something else. The term was applied by Heisenberg to quantum entities such as fundamental particles, as distinguished from 'the things and facts' of ordinary experience.

Presentation (p35): If SX is the associated physical object (q.v.) of the corporeal object X, then X is said to be the presentation of SX.

Prima materia (p88): Matter conceived as bereft of all formal determinations.

Reification (p46): The act by which we clothe physical or mathematical entities with imaginary forms and thereby in a way 'corporealize' these entities.

Specification (p52): The empirical process by which a physical entity is defined or determined to some degree.

State vector (p62): The mathematical entity which represents the state of a physical system in the formalism of quantum theory.

State vector collapse (p65): A discontinuous or instantaneous change in the quantum mechanical representation of a physical system resulting from an actual measurement. The term is frequently applied as well to the corresponding instantaneous change in the physical system itself.

Subcorporeal (p35): A physical object which is the associated object (q.v.) SX of a corporeal object X. Subcorporeal objects, thus, are none other than the physical entities that are identified with a perceptible object according to the usual interpretation of physics.

SX (p34): The associated physical object (q.v.) of a corporeal object X.

Transcorporeal (p35): A physical entity that is not subcorporeal (q.v.). Fundamental particles as well as 'small' atomic aggregates are transcorporeal.

Vertical causality (p114): A mode of causation which does not act from past to future by way of a temporal process, but acts directly or 'instantaneously'.

Yang (p99): The formal or essential aspect of a thing.

Yin (p99) The material aspect of a thing. Like the Aristotelian conceptions of matter and form, or of potency and act, the terms 'yin' and 'yang' need to be understood in conjunction.

Yin-yang (p99): The familiar Taoist figure depicting the interpenetration of a black and a white field. The *yin-yang* could be termed the icon par excellence of complementarity in its most universal and profound sense.

INDEX OF NAMES

www.ingramcontent.com/pod-product-compliance
Lightning Source LLC
Chambersburg PA
CBHW060526130626
46553CB00002B/662